ELEMENTS OF A
PHILOSOPHY OF TECHNOLOGY

CARY WOLFE, *Series Editor*

(continued on page 310)

ELEMENTS OF
A PHILOSOPHY OF
TECHNOLOGY

On the Evolutionary History of Culture

ERNST KAPP

Edited by Jeffrey West Kirkwood and Leif Weatherby
Translated by Lauren K. Wolfe
Afterword by Siegfried Zielinski

posthumanities 47

UNIVERSITY OF MINNESOTA PRESS
MINNEAPOLIS · LONDON

Published by the University of Minnesota Press
111 Third Avenue South, Suite 290
Minneapolis, MN 55401-2520
http://www.upress.umn.edu

Printed in the United States of America on acid-free paper

The University of Minnesota is an equal-opportunity educator and employer.

25 24 23 22 21 20 19 18   10 9 8 7 6 5 4 3 2 1

Library of Congress Cataloging-in-Publication Data

Names: Kapp, Ernst, 1808–1896. | Kirkwood, Jeffrey West, editor. | Weatherby, Leif, editor. | Wolfe, Lauren K., translator.
Title: Elements of a philosophy of technology : on the evolutionary history of culture / Ernst Kapp ;
    edited by Jeffrey West Kirkwood and Leif Weatherby ; translated by Lauren K. Wolfe ; afterword by Siegfried Zielinski.
Other titles: Grundlinien einer Philosophie der Technik. English
Description: Minneapolis : University of Minnesota Press, [2018] | Series: Posthumanities 47 |
    Translation of: Grundlinien einer Philosophie der Technik. | Includes bibliographical references and index. |
Identifiers: LCCN 2018008957 (print) | ISBN 978-1-5179-0225-4 (hc) | ISBN 978-1-5179-0226-1 (pb)
Subjects: LCSH: Technology—Philosophy. | Inventions—Social aspects. | Human mechanics—Philosophy.
Classification: LCC T14 .K213 2018 (print) | DDC 601—dc23
LC record available at https://lccn.loc.gov/2018008957

All of human history, upon close scrutiny,
ultimately resolves into the history of the invention
of better tools.

—EDMUND REITLINGER

Contents

The Culture of Operations
Ernst Kapp's Philosophy of Technology

JEFFREY WEST KIRKWOOD AND LEIF WEATHERBY

THE FIRST PHILOSOPHY OF TECHNOLOGY

In December 2016, Graphcore, a company based in Bristol, England, released computational graphs that were widely received in popular media as "scans" of machine brains, the "neural nets" at the heart of artificial intelligence research in the emergent field of machine learning. The promise of the images was immense. If we suppose that machines are in fact capable of "learning," then, by visualizing the structures and patterns of their operations, we stand to find out something about how the human brain works as well. The materialism of neuroscientific research is taken here to a logical extreme. Not only is thought treated as reducible to purely physical mechanisms, but the technologies believed to replicate or mirror them become a crucial source of insight about how they actually work. Here technology does not simply reproduce or improve human functions, but rather creates the framework for understanding them. Where the human brain remains mysterious, the supposed "machine brain" seems to offer clarity. This is a trend that appears to be gathering momentum: a group at Google has recently published a paper seeking to map the activity of neural nets onto natural language, making AI "interpretable."[1] The effort has been dubbed "artificial neuroscience."[2]

The assumptions underlying such developments are not limited to questions of the digital age. They demonstrate the persistence of a much older theory of technology first articulated in the idea "organ projection," the key concept in Ernst Kapp's 1877 magnum opus, *Grundlinien einer Philosophie der Technik (Elements of a Philosophy of Technology)*. Like the "AI brains" understood through a belief in their isomorphism with human brains, technologies were, for Kapp, projections of human organs. Organs, the operational parts of animal bodies, were the initial blueprints for tools, for machines, and even for institutions and the state. But more importantly, they were also the lens through which the functions of the body and with it the whole universe of human activities became interpretable. Unconsciously, the finger is projected into the stylus, the arm into the ax, and the nervous system into the telegraphic network, allowing humans to emerge as cultural beings—that is to say, to emerge at all.

For Kapp the organ-projection thesis is not a stand-alone theory of technology, but the springboard into an entire theory of culture as operation.[3] What we make makes us, conditions our insights into the world, and even shapes knowledge of bodies. However, we arrive late to this interpretation, which remains irrevocably embedded in the play of operation and sign, function and being. *Elements* was published just as the word *technology* began to crystallize as an abstract noun, and the book was an idiosyncratic but important contribution to that process, even as it insisted on an understanding of technology as something more akin to the contemporary theoretical term "cultural technique." Kapp's book places cultural objects in contact with technological operations from their very inception, which is perhaps what explains his oblique influence on the psychoanalytic, media-theoretical tradition that runs from Sigmund Freud to Friedrich Kittler. Deploying Hegelian methods, he also joins the ranks of those philosophers who consider technology a vocation of the modern world as such. Kapp writes what might well be the first theory of culture as technologically conditioned, as *operation*. The very category of the human, persistent as it is in Kapp's book, vanishes into these operations.

The human, Kapp asserts, is the limit of any possible philosophy. This is a claim that would give some scholars the pretext for mistakenly dismissing his work as anthropocentric. Kapp seems to exacerbate the situation by claiming that "the inalienable form of all cognition is and has always been based on a dualism,"[4] and that somehow only the "whole human" can be the measure by which "all" is taken. Kapp writes—in explicit imitation of Hegel—"the human being who in thinking proceeds outward from himself is the requisite condition for the human being returning to himself."[5] Projecting its organs into technology, the human first learns of the operations of its own body and mind belatedly, in a dialectical process. Thus, for the human to take the measure of itself is not only to recognize the difference between subject and object, but also to *have realized* that difference. What appears humanistic in Kapp is a disjointed humanism at best, and something much closer to what in recent years has been called posthumanism in its recognition of the dialectic between the physical and ideal, information and materiality, and subject and object.[6]

At first glance Kapp's notion of organ projection hardly seems a good candidate for a posthumanism, as "projection" appears to be a firmly human category. We might even call it an anthropomorphic one, since it defines the technological as a process of casting forms out from the physiological contours of *Homo sapiens*. What we miss in such a reading, however, is that technologies *in the way that they operate* are what make the human form comprehensible in the first place. For instance, the significance of the electrical telegraphic network that unified the informational space of the American continent during the second half of the nineteenth century was not that it morphologically replicated the nervous system. Rather its function demonstrated the unity of the physiological apparatus through the systematicity of its transmissions, not its fundamental physical or metaphysical coherence. Object and operation were fused in technologies in ways that returned the insights about their functions to the human bodies to which they were ultimately tied.

Kapp's humanism emerges as a posthumanism because technologies provide the epistemological precondition for the concept of the human by demonstrating the inseparability of *use* and *being*, or one could also say, *information* and *matter*. His understanding of technology preserves the full force of the German term *Technik*, which has found imperfect translation as "technics," "techniques," and "technology," suggesting that the essence of technologies is in how they work. This line of thinking creates an important precedent for theories of posthumanism and cultural techniques *(Kulturtechnik)*, both of which have insisted on the profound entanglements of operation and being in questions of what humans and technologies are. The result is Kapp's invention, in the last third of the book, of a new concept of quantity—*Inzahl*, translated here as "innumeration"—which takes measure of the organism through the lens of the machine, itself projected from the organism. Philosophy of technology, in turn, takes the measure of the human who would write it.

If there is a single organizing force that creates coherence among the panoply of theoretical commitments that constitute posthumanism, it resides in the exteriorization of the conditions for knowledge—so completely in fact that in certain cases the question of knowledge itself, which at some level presupposes a "knower," is demoted to near irrelevance. The familiar ontologies of humanistic discourse that separate humans from animals, and subjects from objects, along with the Kantian conditions for knowledge that parse the world of things according to the possibility for their being cognized by a transcendental subject, have leaked from their hermetic containers into a vast plenum. For all his talk of "the human," it is Kapp who punctured the containers.

Humans are distinguished by an embodied inhabitance and reflexive recognition of difference, which is ultimately what Kapp refers to as "self-consciousness." But there is nothing transcendental about consciousness for Kapp. It is not an ontological human property. There is no absolute horizon of Being nor *Dasein* in the Heideggerian sense, against which ontic differences appear. The position of *Dasein*, if one were to import such terms, for Kapp was the secondary effect of the orienting force of technics. Beings and Being itself are ontic for Kapp in that they can only be said to exist once organ projection has initiated the dialectical process of signifying differences. Humans are different on account of their recognition of difference, not their self-identity, which is what makes Kapp's book so prescient with regard to American posthumanism and German media theory. As Bernhard Siegert writes of cultural techniques on precisely these questions:

> The concept of cultural techniques clearly and unequivocally repudiates the ontology of philosophical concepts. Humans *as such* do not exist independently of cultural techniques of hominization, time *as such* does not exist independently of cultural techniques of spatial control. This does not mean that the theory of cultural techniques is anti-ontological; rather, it moves ontology into the domain of ontic operations.[7]

"Self-consciousness" then is shorthand for technologically based, material differentiations whose differences can only come to signify through technics. It is a kind of recursive function that, wherever difference is detected, makes further operations of differentiation possible while re-inserting the original distinctions into the system. This connects Kapp's work not only to media-theoretical evolutions in cultural techniques associated with scholars such as Hans Ulrich Gumbrecht, Berhard Sieg-ert, and Thomas Macho but also to a tradition of systems-theoretical and posthumanistic thought ranging from Niklas Luhmann to N. Katherine Hayles.

On this point, Kapp's *technics* calls to mind posthumanistic deployments of the term "embodiment." As Mark Hansen writes, "technologies effec-tively *force* us to experience changes in our material environment that are no longer thematizable in representational terms, that can only be lived through at the level of our embodiment."[8] Organ projection presents a similarly difficult situation, in that it is the ground against which the dif-ference from the "environment," and therefore all representation, arises in the first place. This offers no suitable language to describe how the alternation of projection locates humans beings in the bodies in which they have always dwelled but needed tools to recognize, represent, and conceptualize.[9] For scholars like Hayles, analysis of bodies as loci of the cultural discourses centers on "embodiment," which is "in continual inter-action with constructions of the body," and "always already imbricated" with "culture."[10] Such analysis surreptitiously upsets efforts to essentialize the body by acknowledging a certain material organizability that refuses to clearly delineate between matter and the discourses that make sense of it. This demystification offers solid grounds for enlarging the histori-cal frame of posthumanistic theoretical commitments to a time before the cybernetic theories from which they normally draw to see Kapp's relevance to computer-age thinking. For instance, in Donna Haraway's watershed essay on the "cyborg," long canonized as a key posthumanis-tic text, she moves away from stable categories and origin stories toward a recognition that genetics, gender, sex, sociality, and technology are all equally configured as "code."[11] Seizing control of the code reveals total mutability that finds no essential difference between "organism and the machine," which exist as a hybrid.[12] Kapp's understanding of technics as the core of distinctions such as "human" and "machine" reveals this in-separability and hybridity long before the Apple I or the Human Genome Project.

Where the emphasis on "materiality" has been fruitful in the recent theories of media it has also sometimes forfeited the operational nature of technology that was built into the older term *technics* used by Kapp. For him, technology was operational first and object second. Organs were projected into tools, but, as technics rather than technologies, their very use opened the world of culture as a world of crucial distinctions, between the open and closed, built and unbuilt, or "the raw and the cooked."[13] The "cultural technique" of the door, as Siegert has argued, is not just a dead mechanism, but operates "the primordial difference of architecture," between "inside and outside," regulating the distinction that allows two spaces to signify.[14] Its basic mechanism is at once symbolic and material, technological and cultural. This was an inseparability that was at the center of Kapp's theory of "organ projection," motivating his notion of technics, and predicting many of the media-theoretical and posthumanistic developments that would come after him.

For more than a century, the philosophy of technology has opened territories of thinking about questions of representation, material culture, human bodies, and epistemological, even metaphysical categories that are now revealing themselves in laboratories and start-ups. But "artificial neuroscience," even as it follows the pattern of scientific inquiry Kapp laid out, is not likely to yield insights about representation and logic all by itself. For that we need a philosophy of technology, or better still, of "technics" as Ernst Kapp conceived of it.

What is "technology"? As Leo Marx and Eric Schatzberg have exhaustively documented, the English term *technology* and the German *Technologie* signified "the study of production" deep into the nineteenth century. L. Marx's example is instructive: the Massachusetts Institute of Technology, founded in 1861, has as its mission the study of production, not of the artifacts of that production. Proto-progressivist celebrations of the new machines were abundant at midcentury in the United States. Marx isolates the singular influence of Secretary of State Daniel Webster in the formation of the notion of a collective phenomenon that would later be called "technology." Webster's notion took a prominent place in public debate in the 1850s, when Kapp happened to be in exile in Texas. The way we commonly use the term today omits the "study" (-logy) contained in the Greek etymology, designating the sum of artifacts and machines. This usage dates to the end of the nineteenth century, for most of which "technology" was the "study of production," not its product. A word was needed to describe the new, machinic aspect of what Hegel had called objective spirit, but it was lacking. As Schatzberg has detailed, it was the German *Technik* that was translated into "technology" in the United States in the late nineteenth century. Only a few decades later, Heidegger could blithely refer to the "question concerning technology" and bemoan a reified sense of the word that concealed its metaphysical vocation. Kapp's complex usage, vying for prominence in the term's infancy, did not win the day. It stands, however, as a metaphysical intervention comparable to Heidegger's. It was the first—and perhaps until today the only—Hegelian philosophy of technology proper, and it forged a working connection between operation and being that brought a whole universe of discourses on technology into conversation with one another.

Kapp has become a familiar, if neglected, figure in the history of media and technology. Where he is cited at all, it is generally as a straw man for weak anthropocentric ideas about the body as the source for technology, or as a minor predecessor to the work of more important theorists. Most notably, there is agreement in the fields of media theory and the history of technology about Kapp's position as the first to introduce a philosophy of technology and also about his role as the accidental forerunner to Marshall McLuhan's idea of a medium as any "extension of ourselves."[15] While it has acted as a bulwark against Kapp's total erasure from the record, this conflation of "projection" with "extension," first made by Friedrich Kittler, mutes Kapp's theoretical power and assigns him to a subordinate role among titans in the history of technology.[16] As a part of Kittler's misdiagnosis of organ projection, Kapp has also been linked to Freud's famous dictum that through technology the human becomes a "prosthetic god." (Despite Kittler's claim that Freud "never once read" Kapp, there is good reason to think that he did.[17] Not only did Freud have the book in his personal library, but it was among the collection of books he selected to bring with him to London when he fled Vienna.[18]) Historical speculations notwithstanding, Kapp's work lays important groundwork for the mobilization and interpretation of core psychoanalytic concepts—such as projection, the unconscious, and the mirror stage—by later media theorists. Similarly, *Elements* has also been viewed as a predecessor to Arnold Gehlen's description of technologies and institutions as "ersatz organs" compensating for the "unfinished" or "lacking" human—a lack that defines humans, paradoxically, precisely through what they are not.

While the resonance of Kapp's work with a more established theoretical canon is remarkable, there is also something strange about his assimilation into these anthropologies. Kapp does not describe a "relationship" of humans to technology. As we will see, he cannot take the modern sense of the word *technology* for granted. It is better to say that technology, for Kapp, *is* the relation between humans and the world and the precondition for the emergence of both. Technology is no regional ontology; it is instead the fundamental process that gives rise to the concepts and the artifacts that we use to define it, always retroactively. This, we would propose, puts Kapp into a class of philosophers of technology, such as Martin Heidegger and Gilbert Simondon, who think of their object as primary, as intimately involved in the production of the concepts and symbols used to talk about it. In this sense, the artifacts and machines so often grouped imprecisely under the static heading of "technology" become the generative starting point for analysis. But they cannot be taken as static, as the concept "organ projection" demonstrates.

ORGAN PROJECTION

If tools, machines, and systems are all projected from organs, then we only discover these organs after the fact. What sounds like a neat linear process—"organ projection"—turns out to be the matrix of all cultural activity. The subtitle of Kapp's book—*On the Evolutionary History of Culture*—designates the dynamic generation of material and symbolic distinctions. Organ projection is the plastic procedure in which all relations of the material and the symbolic are generated. This encompassing definition makes technology all too human, but it dislocates the very humanism that forms its premise.

The book begins with a robust defense of "anthropometrism," the doctrine that humans are the measure of all things. The wide-ranging first chapter is followed by a programmatic second devoted to the notion of organ projection. The projection of form, as in the case of the ax and the arm, is not the only type of projection, we learn. Technology, Kapp tells us, is in fact an *externalization* of an image *(Bild)* that produces an artifact. The artifact is often characterized as an *Abbild*, which means both "image" and "derivative." This in turn means that there is a *Vorbild*—

which means both "before-image" and "prototype"—and thus an "after-image" *(Nachbild)* that reveals the functional and morphological likeness between technologies and the organs from which they were projected. Initially, it seems that the image is the arm, the after-image the ax, and the prototypal image a kind of preconscious, imagistic activity that leads to the production of that ax.[19] Framed as a causal sequence, preconscious activity causes humans to make copies of their organs as tools that then retain the traces of their origins. But things are not so simple, and this introduction is meant to show that such interpretations are not only misguided but have kept Kapp from the status of first-rank thinker of technology, and so deprived us of a major paradigm in the ongoing confrontation between technics and thought.

The language of images Kapp relies on heavily throughout the entire work is not incidental to the work's interpretation. His frequent play on various forms of *Bild* establishes an optical relationship between mechanisms in the real that are otherwise discrete, separate, and in a certain respect incoherent. Let us clarify by analogy. Think of an elaborate arrangement of lenses in a dark room where little light has ever penetrated. Imagine someone entering the room who is entirely unacquainted with optical devices. The discovery of the room is rather unremarkable at first. It is not clear what the lenses are, what they do, or whether they stand in any relation to one another. Suddenly, however, a light floods the room, passing through the lenses to reveal a brilliant image projected onto a wall. In an instant, the image on the wall reveals that the network of lenses is a single system. In a certain way, they were always a system, but without first seeing the image, it would be impossible to view them as a part of an autonomous whole. The unity of the image is an "after-image" of the sequence of lenses, which allows one to look back at them as a "prototypal image" for the image on the wall. In this way the image on the wall is a "likeness" *(Abbild)* in that it captures something essential about how the system of lenses works. What is important here, however, is not tracing the causal order or the ontological status of the images, but the epistemological, operational fact that the image and the lenses only make sense on their own when first viewed in relation to one another. Prior to seeing them act in concert to produce the image, one lacks the possibility for understanding the order of the lenses in total, the individual functions of the separate lenses, and the image on the wall as the product of the arrangement. The light in the room is the first moment of consciousness, and it shows us, if we look carefully, how we have externalized our organs through a complex projection of their functions, forms, and operational protocols. In Kapp's notion of organ projection, we come late to the illuminated scene. The only difference is that the image-making apparatus is internal to our bodies and is "projected" out as a material externalization.

Kapp has too often been read as defending a linear ontology of technology that proceeds from pre-harmonized organ-patterns like the arm and its sinews into tools like the American ax. Reading Kapp, we learn two things almost immediately that undermine this linear picture. The first is that the process of externalization is *unconscious* (chapter 10 of Kapp's book is devoted to this), meaning that the "image" is realized only after the concrete artifact, in this case the ax. The process of fabrication is retroactively determined and thus forces knowledge to appear only after its productive capacity. "Form is given to the tool long before the formulation of the law that is only then later experienced and acknowledged as an unconscious endowment accompanying the form itself."[20] For Kapp, technology is not the product of projection but the process itself, not the ax but the continuum between arm, ax, and image. Whatever process selects the "image" from the organ to be projected, it is only through the externalization that the image comes to consciousness. Technology becomes a mediator of knowledge in general, but always after the fact. This temporal loop undermines every reading of Kapp that simplifies organ projection into a linear (and always implicitly conscious) process.

The second factor that undermines the linear understanding of technologies as simple organ copies is that the organs that are allegedly copied vary fundamentally from each other. In the ax, the arm's form is copied. In the railway system, it is the vascular network that is reproduced, as is the brain's structure in the telegraph system. But as we will see, not all organs are copied in form, a fact made apparent in Kapp's heavy reliance on the work of the German engineer Franz Reuleaux, who analyzed machines kinematically according to abstract geometries of motion rather than the concrete interactions of individual parts. In kinematics, a machine was understood as a whole—as an image of the full range of possible movements as determined by pairings of mechanisms.

This view finds no simple projection, but instead a principle of engineering that, when returned to the body, makes sense of the relationships between the parts and the whole that the body on its own was insufficient for producing. The "image" that is here "copied" does not exist in prototype and was therefore not a representation or a concept before the projection is realized. Organ projection becomes the dynamic milieu from which both machines and self-consciousness emerge. This distinction, it turns out, is at the root of all other distinctions—difference itself is technological.

"Technology" is the matrix that generates the very terms we use to characterize not only artifacts and machines but also bodies and minds. The term *organ*, from the Greek *organon*, "tool," has always been at the bleeding edge of epistemology. Kapp uses Aristotle's famous maxim "the hand is the tool of tools [*organon pro organon*]" in chapter 3, for example, playing on the equivocation between organ and tool. But even if we accept a broad definition of *organ* that includes organisms as well as language, what should we make of projection? What logic is invoked in this only partly conscious but anthropologically central process?

Chapter 7 presents us with the crucial transition from the projection of the organ's form and tool-function to the machine. From this point it will no longer be the phenotype of the human that determines the technological "image" or artifact, but instead the *processes* of "organs" of all kinds, including bodily systems. Cardinal among them is the vascular system, comparable to the steam engine, which is "known, admired, and in use the world over . . . truly a 'universal all-purpose machine.'"[21] The comparison, as Kapp exhaustively documents, is anything but original. One thinks immediately of Karl Marx's comparison of the vascular system to the "instruments of labor,"[22] or of Kapp's own source, Hermann von Helmholtz, whose influential treatise *On the Interaction of Natural Forces* belabors the analogy between automata and animals to demonstrate the universality of the second law of thermodynamics.[23] Kapp extends the analogy he finds to encompass the emergent global transportation system: "Rail lines united with steamship lines to form a closed system, the network of transport arteries through which circulate humanity's means of subsistence—a likeness of the blood vessel network in the organism."[24] But he warns us that the inventors James Watt and Robert Stephenson in no way allowed "their own bodies to stipulate for them the law and the norm for the mechanical structure of their machines," although "the coordination between them is so exact that this affinity between the organic prototypal image and its mechanical after-image serves as a considerable source of evidence for even the most respected representatives and interpreters of knowledge concerning human beings."[25] The shift from tool to machine expands the notion of the "organ" in the phrase "organ projection" beyond any obvious visible pattern. No final form of the body can stand at the origin of technology, since any process in that body might adumbrate a projection, which will then crystallize into an image only when the technology has already resulted, in turn first giving rise to a new moment in the science of physiology. Between machine science and biology stood a conditioning discipline called "philosophy."

The philosophy of technology came into being as a way to struggle with the largest-scale transformations of nineteenth-century Europe, both political and scientific. For Kapp, that meant it had to act as a revision of the philosophies that had already grasped this moment—those of Hegel and his skeptical enthusiast, Ludwig Feuerbach. Feuerbach supplied part of the projection-notion, while Hegel offered the logic and the metaphysics. But neither had witnessed the creep of machines like the steam engine across the globe. A philosophical update was in order.

The steam engine's appearance of autonomy, partly based on the function of the governor apparatus that controls its speed using mechanical feedback, does not so much "imitate" organs as it reveals them and their functional wholes—organisms—for the first time. Kapp writes:

> What instills in us such great admiration for the steam engine is not any of its technical particulars—not, for instance, the reproduction of an organic articulated joint using rotating metal surfaces lubricated with oil; not the screws, arms, hammers, and pistons. Rather, it is the feeding of the machine, the conversion of combustible materials into heat and motion—in short, the curiously daemonic appearance of a capacity for self-motivated work. Here we are reminded of the higher provenance of the machine in the human being, whose hand built the monstrosity and set it free to compete with storm and wind and waves. Every scrutinizing glance cast upon it contributes to establishing the truth of Feuerbach's proposition that *the object of any subject is nothing else than the subject's own nature, taken objectively.*[26]

Ludwig Feuerbach's critique of religion is here shifted into the techno-cultural process. Feuerbach had given the spark to the anti-religious political impulse of the Left Hegelians, claiming that "religion . . . [is] *the unmediated object, unmediated essence of the human*"²⁷ and nothing else. Religion, Feuerbach is suggesting, is the projection of human essence into theological categories.²⁸ In this projection, however, the sensuous and embodied human is present in the image-making: "The essential difference between religion and philosophy is the *image*. . . . Whoever takes the image away from religion, takes away its subject-matter [*Sache*], and has only the *caput mortuum* in his hands. The image is *as image* matter [*Sache*]."²⁹ What matters in projection is not the resulting image but the projection itself. Kapp responds with his own alternative:

> It should be self-evident that what is meant by "image" here is not the allegorical figure that we have mentioned at several points already. That is, the "image" is not a pure simile that puts at one's disposal an apparently unlimited number of possible comparisons; rather, it is the concrete image [*Sachbild*] and reflection of the projection, which exists uniquely.³⁰

Note that *Sachbild* seems stripped directly from Feuerbach: the image *as object* is not an analogy after comparable images have been produced. Instead, the process of producing those images—organ projection—is both embodied and material, progressive and dynamic. Marx was among the first to recognize the "monstrous" autonomy of machines, especially those at work in the early automatic factory.³¹ Kapp, who has too often been construed as limiting technology to the "image" of the human, finds a deeper ontological process that first imprints the technological cosmos and by extension the natural world itself. The dialectic of science and technology takes on a properly Hegelian hue.

Just as Hegel's project does not seek to explain the absolute origin of life, Kapp is not concerned with a time prior to history—that is, with a time prior to technology. Even where he mentions evolutionary theorists, for instance, he is quick to qualify his references as concerned with the processes of evolution rather than their origins. It is important to see organ projection not as a topic within history but as the epistemological operation by which history becomes possible at all. The facts of human history are not new; what is new is his "treatment of them" from "the point of view of projection," which "is applied" in his book for the "first time."[32] Kapp's novel theoretical optic of organ projection offers the "historical and cultural foundations of epistemology [*Erkenntnistheorie*] in general" as inseparable from human tool use.[33] Any notion of a history prior to what is immediately present is an effect of external technologies that show humans to themselves for the first time from the outside. Everything prior to this point is, strictly speaking, ahistorical.

Kapp draws openly on the evolutionary theories of Lamarck, Darwin, and Haeckel, acknowledging questions of descent as "*the* scientific of issue of the day."[34] However, where "the origin of species" was concerned, Kapp was only occupied with the epistemological grounds upon which humans as cultural and historical beings came to recognize themselves as such. Without technology there was no evolution, although not in the sense that the biological and environmental factors involved in mutation and selection did not exist prior to the implements of the human hand. Rather, evolution as an explanatory apparatus for the trajectory and forms of life, whose scale was necessarily larger than the scope of any single living being, could only arise through technologies that allowed humans to imagine themselves abstractly as part of a history they could not have witnessed. In this way, technology for Kapp did not simply augment or extend human faculties. One likely source for the erroneous reception of organ projection as a mode of organ extension was his insertion into the evolutionary discourses he cites, which saw the emergence of tools as a moment on a longer evolutionary timeline rather than as a precondition for all knowledge production. The evolutionary role of tools and technologies for the social Darwinist Herbert Spencer, for instance, was much closer to McLuhan's extension thesis, where he identified "supplementary senses" and "supplementary limbs" that marked the artificial conditions necessary for the development of "higher forms" that characterized humans.[35] Spencer even called "all observing instruments . . . *extensions of the senses*" that produced fundamental changes in the ways that humans related to their environment.[36] Yet where Kapp represents a departure from the thinking of his contemporaries is in his recognition that such a distinction was the result of technologies in the first place. Part of the brilliance of Kapp's position that has gone so woefully unnoticed is his total commitment to an understanding of humanity's position as the unwitting author of history, which could not be written but for the pen from its own hand. Although a reciprocal dependency between "artificial senses" and "artificial limbs" bears some similarity to Kapp's idea of organ projection, reading his work as a part of evolutionary theory has helped obscure much of the nuance of his position and to flatten the dialectical complexities of his epistemology.

For Kapp humanity was not a "rung" in a linear evolutionary sequence, and technology did not simply extend human faculties with increased speed, reach, or proficiency. Tools and physiology made symbolic activity possible, but symbols allowed the textures of the material to be recognized, articulated, represented, and preserved as culture and language. The passage between these two realms was technics, which, as both protocol and materiality, moved ceaselessly between questions of being, doing, and knowing. As a result, Kapp's techno-dialectical Hegelianism allowed for no animals without humans, no humans without culture, no culture without technology, and ultimately no meaning without materiality. Through a proliferation of dynamic oppositions, humans, and with them language and culture, offered the framework according to which the difference between humans and the animals from which they evolved became visible. For example, the "parallelism"[37] between the electrical telegraph and the human nervous system was a near commonplace assumption in psychophysical research by the prominent nineteenth-century scientists Kapp invokes, such as Emil du Bois-Reymond, Wilhelm Wundt, and Johann Nepomuk Czermak. If the force of his observation were only the comparison of the two systems, suggesting that telegraphic networks were modeled on nerves or that mere "symbols and similes are at play," he would have added very little to the existing literature on which he was commenting.[38] Instead, however, he claims in no uncertain terms that while the material configuration of the body is unconsciously projected into the structure and function of technologies, "the mechanism acts as a means of disclosure and understanding of the organism," which is "the real essence of organ projection."[39] Telegraphic networks were based on the nervous system, but, conversely, they also helped us understand it.

Organ projection as operation first made sense of the human body, its relation to the world, and consequently its relation to the very technologies that it projected. The shape and function of the "tools and machines" that propelled the "mechanistic worldview" of nineteenth-century empirical psychology could not be divorced from the bodily form that they came to regulate.[40] They instituted differences that led to "an orientation toward the organic world and a power that aids it conceptually."[41] In other words, technology first revealed the difference between humans and their environment before helping them to see where they were located within it. The breathless PR of Graphcore's metaphorical "brain scans" follows this logic precisely. The brain projects not a telegraph but another "brain," and any image of that brain must be reduced back to the language and instruments of neuroscience. Kapp predicts that science will follow technology in this way, something we almost take for granted in the twenty-first century.

THE QUESTION CONCERNING TECHNOLOGY, FOR HEGELIANS

Kapp was a singular Hegelian. Rather than operating within Hegel's system, or adapting it to some particular area of research, he synthesized Hegel's political and metaphysical speculation with the fast advance of technology he saw during his lifetime. Technology was spirit, both objectively and in an absolute sense. It provided the substance of philosophy and was the principle of that philosophy's exposition.

The title of Kapp's work is modeled on Hegel's *Elements of a Philosophy of Right* (1820). It was that work which gave the political edge to Hegel's philosophy that spawned movements of Right and Left Hegelians after his death in 1831. In the preface, Hegel outlined a metaphysical understanding of politics, writing, "That which is rational is actual; and what is actual, that is rational [*Was vernünftig ist, das ist wirklich; und was wirklich ist, das ist vernünftig*]."[42] This enigmatic statement touched off one of the great interpretive wars in philosophical history. Those who wanted to defend the Prussian state took Hegel to mean that the actually existing social order was rational—all that remained was to refine and maintain it. The Left Hegelians, among them Moses Hess, Bruno Bauer, and Ludwig Feuerbach, understood the sentence actively: what is real must conform, or *be made to conform*, to the rational.[43] It was in that spirit that many of them joined the socialist uprisings that swept Europe in 1848, upending governments and—in their failure—sending countless young radicals into exile. Kapp was among those who fled the German monarchies,[44] settling in Sisterdale, a colony of German immigrants in

Texas. It was here that he encountered the American ax, and with it a cocktail of agrarianism and pragmatic technological orientation where tool and machine were integrated with human and animal, paradigmatically in the plow.[45] Here Kapp led an abolitionist group called the Free Society (Freier Verein) in the confrontational years that led to the Civil War. And somehow, he also managed to bring a complete edition of Hegel's works with him.[46] Agricultural life, offering water cures at a spa he operated, and participating in a socialist experiment took up nearly two decades of Kapp's life. He returned to Germany in 1865 and took medical advice not to risk the trip back the United States. *Elements*, published a decade later, reads as a synthesis of political, practical, and intellectual experience at the edge of world history.

Kapp merged his experiences on the American continent with his earlier Hegelianism. He was among the first to include technology in Hegel's notion of "objective spirit." Hegel's system distinguishes between three kinds of spirit *(Geist)*: subjective, objective, and absolute. Subjective spirit, the travails of which are familiar from the *Phenomenology of Spirit*, is the object of psychology, the development of consciousness and self-consciousness as the *for itself*. In this respect subjective spirit is a one-sided expression of the absolute "idea." So too objective spirit, which expresses the other side, the "absolute idea, but only as *in itself*."[47] But if *spirit* is "in itself," this means it is *objectified*—we could even say projected— and made into, as Hegel puts it, "the anthropological side of particular needs" and into "external things which are for consciousness."[48] In other words, objective spirit is spirit *externalized* in its aspect as *for* the externalizing spirit. This encompasses anything that spirit produces and then uses, needs, or is influenced by—including institutions of all kinds, governments, societies, and markets. These institutions, Hegel writes, are "freedom, formed into the reality of a world" such that it "obtains the *form of necessity*."[49] These are generally the media in which subjective spirits encounter one another, in which the infinite variety of a conflict of wills emerges. Kapp follows Hegel closely on this point when he writes:

> The outside world comprises myriad things that, though nature did supply the material, are less the work of nature than the work of the human hand. As artifacts, by contrast with natural products, these make up the

content of the world of culture. The "outside world" for the human being is therefore composed of both natural and man-made objects.[50]

The notion of objective spirit played a central role in the disciplinary developments in the human sciences that occurred in Germany around 1900. Georg Simmel and his fellow early German sociologists used this concept to define the realm of their inquiry: the "form of necessity" unfolded as the proper laws of the social.[51] Wilhelm Dilthey, whose hermeneutics first gave theoretical form to the modern humanities, altered the concept slightly to encompass symbol making and its technical transmission in writing ("the written remains of human existence")[52] in order to aver the substantial continuity of historical interpretation or "understanding" (Verstehen) with the past it aimed to understand. In other words, objective spirit had been repurposed to form the basis of both the social sciences and the humanities by 1900.

Each of these uses borders on the direct association of objective spirit with artifacts, machines, and inscription processes. The title of Kapp's opus is a claim, easily recognizable to the academics he needed to impress once back in Germany after 1865: the form of necessity that attaches to the world of convention is dominated by something called Technik. The dialectic of nature and spirit determines a world that expands the realm of objective spirit in relation to its components (spirit and nature), eventually determining them retroactively and providing the basis for the interpretation of both.

Kapp expands technology not only into objective spirit but even into "absolute spirit," which Hegel defines as "identity" and in turn combines substance and judgment.[53] This dialectically whole spirit has its own history, proceeding through the realm of the senses in art, through the heart (Gemüt) in communitarian religious spaces, and into itself as its own medium in philosophy. Absolute spirit is the realm of human endeavor where cognition is embedded in the external in such a way that they are harmoniously united. And this is precisely how Kapp defines technology.

We can see this most clearly when the "organ" projected becomes entangled with the mind. Chapter 8 is devoted to the projection of the nervous system into telegraphy. The cables laid throughout the Western world in the second half of the nineteenth century seemed a good analogy to many for the brain and the central nervous system: patterns of branching wires carrying messages like thoughts or perceptions across the globe. Hegel was dead when the commercial aspect of electrical telegraphy began in the 1830s. But it would confuse the difference between objective and absolute spirit in a way that left a permanent imprint on Kapp, and through him on the philosophy of technology. Kapp writes:

There is no other organ whose character is displayed as clearly in the construction that has been unconsciously formed in its image [nachgebildet] than the nerve cord is in the telegraph cable. Here organ projection celebrates a great triumph. Its principal requirements being: production taking place unconsciously on the model [Muster] of the organic; followed by an encounter in which the original and the copy find themselves, compelled by the logic of analogy; and finally, revealing itself like a light switched on in the conscious mind, the agreement between the organ and the artifactual tool, with the greatest conceivable degree of identity.[54]

Organ projection is confirmed by the growth of telegraphy, which imitates (nachbilden) not only the form of the nerve but, more importantly, its function, in this case the transmission of ideas through material, whether nerves or wires. But engineers like Samuel Morse did not work with this image in mind any more than had Watt and Stephenson with the steam engine. Instead, the completion of the machine and then its systemic spread first allows the discovery of what was invented. The identity is unconscious. Spirit has already externalized itself when it comes to realize the analogy between nerve and wire. The telegraph articulates a point that the ax could not have alone: objective spirit borders on absolute spirit in the expression of the unity of idea and material, here modeled on the brain. This moment is crucial in Kapp's exposition and imparts a Hegelian core to his form of posthumanism. He asks rhetorically: "We have seen the unconscious take part in the emergence of the tool's basic forms and in their nearest modifications. Is it supposed to cease participating when we realize [vergegenwärtigen] that even the highest achievements [Leistungen] of conscious artifaction [Artefaktion] ultimately, in their turn, unconsciously serve the progressive revelation [Offenbarung] of the entire human being as thought-entity [Gedankenwesen]?"[55] The process of organ projection, with its strange temporal loop, cannot be eliminated by the shift from form to function. This holds even

when the function is *thought* or its communication, as in the case of
telegraphy. With this move, Kapp shifts the semantic range of "organ"
to operation more generally. This operation, as the intersection of form
and fuction, *is technology*.[56] The projection process externalizes the "whole
human," as Kapp promised, including the intractable problem of the
embodiment of mind. Technology is objective spirit as absolute spirit.
Cognition, and indeed communication in general, are materialized in
technology. Or perhaps it is better to say that technology is the synthetic
condition from which these elements can first be isolated and represented
as though independent. Technology is, in this sense, genuinely dialectical
in Kapp's work: it is the result that is the premise for its parts, which are
artifacts, machines, and inscription processes on the one hand and symbols, representations, and concepts on the other.

ANTHROPOLOGY AFTER HUMANS

The credit and the blame for the often repeated claim that Kapp was the
unconscious source for the extension/prosthesis theory of technology is
arguably traceable to Kittler's 1999 Berlin lectures, *Optical Media*. Here
Kittler criticized Marshall McLuhan as "a part of a long tradition that can
be traced back to Ernst Kapp and Sigmund Freud, who conceived of an
apparatus as a prosthesis for bodily organs."[57] In *Eine Kulturgeschichte der
Kulturwissenschaft* (A cultural history of cultural studies) Kittler builds a
more extensive historiographical scaffolding for interpreting Kapp than
he did in *Optical Media*, but he still fails to entirely avoid the conflation
of "projection" with "extension." Though Kittler is likely the origin of
many of the persistent misunderstandings about Kapp, one of his central
insights that is overlooked in the rush to affiliate Kapp and McLuhan is
that technology for Kapp was the precondition for any scientific knowledge. Kapp's uncanny appearance in the history of media theory and posthumanism occurs like a return of the repressed. His theory of technics
and organ projection, which has been disavowed as anthropocentrism,
actually prefigured the rethinking of relationships between culture and
media, materiality and signification, and subject and object associated
with the explosion of theory in the 1980s and 1990s. Generally speaking, those theoretical "turns" recovered the scientific materialism hidden
in Freud's psycho-hermeneutics and extended the consequences of technology's critical position in Jacques Lacan's understanding of the real and

the symbolic. And they did so in a refusal of the primacy granted to signification in poststructuralism. In other words, the discourses associated with posthumanism and media theory relied on a hard distinction between the real and the symbolic and between technology and culture, but they inverted existing humanistic theories by moving the site of analysis from signification to the realm of materiality, which was no longer distinct from the information it channeled. Yet this was something that Kapp had undertaken in 1877, having already recognized the inseparability of the categories that would act as the drivers for theory a century later.

The rise of what has been received in the anglophone context as New German Media Theory involved a "war" that pitted "'Cuture' against 'Media.'"[58] Kittler, a leading figure in this conflict, notes that what are called "the arts" only concern themselves with "symbolic relations with the sensory fields they take for granted," while "media relate to the materiality with—and on—which they operate in the Real itself."[59] In the reorganization of analytic priorities accompanying this theoretical upheaval, focus was shifted away from "the representation of meaning" and toward "the conditions of representation"—from "semantics itself to the exterior and material conditions of what constitutes semantics."[60] The scope and nature of the symbolic distinctions from which cultural objects emerged for Kittler were "determined" by media technologies.[61] And Kapp's treatise on technology, the reception of which Kittler himself dictated, may serve to clarify much of the confusion.

Even where Kittler praises McLuhan for privileging the medium over "the message," he rejects the idea that media function as "extensions" of the "human senses" or as "prostheses." Kittler vigorously refused this "old thesis," which "amounted to saying, in the beginning was the body, then came the glasses, then suddenly television, the computer."[62] One could not simply presuppose an existing spectrum of distinctly human activities in the absence of the technologies that structured them, a position that bore deep affinities with Kapp's claim that "tools establish culture, because they participate in the rapport existing between hand and brain."[63] Here tools are not the servants of culture, nor are they mere extensions of the body. They establish a technical "rapport" that bridges between materiality and ideality that was the heart of what would later be named "cultural techniques [Kulturtechniken]." Kapp appears to have set the stage for not only the "war" between media and culture but also for the eventual truce whereby culture and technics became historically inseparable in cultural techniques.

In this sense, Kapp adds a Hegelian lineage to New German Media Theory that represents a significant departure from the way he has usually been read.[64] Ernst Cassirer, for example, made Kapp's program into a partner project for his attempt to include technology in his philosophy of symbolic forms. Acknowledging but rejecting precisely the unconscious element of organ projection, Cassirer writes that Kapp's "fundamental insight" is that "technological activity, in its direction outward, is always also a human avowal of self and represents a medium for him of self-knowledge."[65] It is significant that Cassirer can only arrive at this reading by excising the unconscious process from organ projection. In doing so, he gives up on the belatedness of human cognition with respect to technology, gives up on the dialectical retroactivity of technological synthesis, and abandons any hope of reconstructing the identity between matter and mind that Kapp's theory has in view. But this misreading allows us to see that Kapp's "human" is not a static transcendental, an unmovable phenotype—neither the matter nor the mind remains fixed. Humans, as he argues, in measuring against themselves "the outside world" they had a hand in creating, develop "an ever increasing self-consciousness" that is the mechanism of history.[66]

This line of argumentation begins to clear up the fundamental error that has attended the insertion of Kapp as a supporting figure in the more familiar theoretical tradition of media theory from Freud to McLuhan, namely, the misinterpretation of his claim that humans are "the measure of all things."[67] While it is true that organ projection "endows form to the tool," it is also the "formula for the action of forces" that are put into play.[68] What is called "measure" is intended in the "most comprehensive" sense as a cultural technique, capturing both the "material and ideal" that results from the human-technics dialectic.[69] On its own the body is not a measure, as it lacks an external point of comparison and the operationalized increments to standardize difference. To view the body as a measure of any kind is already to understand the body as subjected to the technologies for which it was the source. The two movements cannot be divorced for Kapp, and in turn the human as "measure" is always already a technological and operational question.

THE INTERPRETATION OF PSYCHOANALYSIS

Elements is a critical missing piece in both the history of psychoanalysis and its reception by subsequent media theorists. There are two areas in Kapp in which this is saliently expressed: the unconscious and organ projection. In the first, Kapp, like Freud, understands the unconscious as a pre-technical materiality rather than a region or place within the psyche. It is the "prototypal image" that can only be interpreted once additional forms of mediation allow one to understand it as the source for later appearances. The unconscious for Kapp, as for psychoanalysis, which saw the unconscious as a storehouse of desires that could not be expressed without mediation, was of a bodily origin. This was deeply connected to the second area of intersection, namely, projection, which offers a techno-optical mechanism of ego formation that predicts the interreliance of the material and the symbolic underlying both Freud's optical conception of the psyche in *The Interpretation of Dreams* and Lacan's mirror stage. In this way, Kapp can be seen as the technological unconscious of psychoanalysis exhumed by media theory, rather than an early evolutionary stage in its emergence.

In *Optical Media,* Kittler places Kapp within a trajectory leading through the passage in *Civilization and Its Discontents* (1930) in which Freud claimed that through "science and technology" man had come close to attaining a "cultural ideal" that had made him into a "prosthetic God."[70] Critical of this formulation, Kittler attacked the "assumption that the subject of all media is naturally the human," which disguised the fact that "technical innovations" progressed "completely independent of individual or even collective bodies of people."[71] The two elements of Kapp's philosophy that conspired to support this reading were the idea that humans were "the measure of all things" and "organ projection," which when paired with the former led to a simplified sense of technology as moving causally from a unified body to amplify its preconstituted faculties. Although a faithful account of these two principles was never offered, in his *Kulturgeschichte* Kittler moves the point of contact between Kapp and Freud to *The Interpretation of Dreams* (1900), insinuating Kapp into the very fiber of psychoanalytic theory.

Much of the work Kapp does in *Elements* that resonates so remarkably with contemporary scholarship in the posthumanities as well as psychoanalysis derives from his engagement with the field of psychophysics, whose fashion in the second half of the nineteenth century allowed it to storm German universities in a coup that replaced many of the chairs of philosophy departments with physiologists—a triumph of the machine over *Geist*.[72] Kapp expressly called figures such as Gustav Fechner, Wilhelm Wundt, Hermann von Helmholtz, and other leaders of empirical inquiry in the nineteenth century the "history" of his own work. For Kapp, "Psychology and physiology [had] been estranged long enough," and it was his aspiration to assist in providing a "physiological psychology" that could overcome the "dualism" and stabilize "the fluctuating concept of personality" by "relocating the concept entirely within self-consciousness."[73] The vehicle for this relocation was the idea of organ projection, which, if one takes seriously the optical nature of the concept, functioned precisely through a dialectical "dualism" between the mechanism and the unified image *(Bild)* it created.[74]

Like Kapp, Freud was trained in the milieu of nineteenth-century science, which relied heavily on mechanical instrumentation for the empirical study of mental processes and analogies between the function of machines and the psyche. This allowed what had been the strictly philosophical terrain of *Geist* (both in the sense of spirit and psychology) prior to Johann Friedrich Herbart to be reconceptualized according to advances in machine technology and thermodynamics that made the psyche into a finite mechanism for managing energetic inputs and outputs.[75] Freud outlined a material model of the psyche in notebooks sent to Wilhelm Fliess in 1895, published posthumously as "Project for a Scientific Psychology."[76] In this pre-psychoanalytic treatise, the psyche was envisioned as a complicated regulatory system directing, storing, and discharging energy that entered it as sensations. No doubt this model was "scientific" insofar as it abided by nineteenth-century expectations that physical systems obey the same principles as the machines that measured them. At the same time, it could not account for the experience of consciousness, which evaded any attempt to locate it in the turning gears of the machine. The solution to this difficulty, which allowed Freud to maintain the mechanical, and thus scientific, character of the psyche as an apparatus while also explaining its curious relation to consciousness, was to liken it to an optical instrument, such as a microscope or a telescope. In his section on Kapp, Kittler cites the famous passage from *The Interpretation of Dreams* in which Freud makes this tentative move from the psyche as mere physical machine to the psyche as symbol machine.[77] But Kapp's work did not merely serve as preparation for this transition in Freud. Kapp had already made this analytical move almost two decades earlier. The heuristic power of the comparison with optical instruments— both in organ projection and *The Interpretation of Dreams*—was that the physical arrangement of the lenses had the effect of producing an image without a physical location within the apparatus. It was an interaction between light, the operations of the device, and the viewer. The image did not reside in the perspectival tube, the lens, or the eye. Optical images were "virtual" in that they depended on the material focal properties of the instruments but were not reducible to them.[78] For Kittler this is the key to psychoanalysis, whose various and sophisticated topologies of the psychological economy of symbols emerged from and never fully shook off their physiological beginnings. Egos could not exist without brains and cultural symbols could not be preserved and transmitted without communication and storage media. Freud's "radicalization of technical Organprojection" was in a sense the birth of psychoanalysis.[79] What Kapp outlined was a link between technical media and symbolic operations that were the historical and conceptual foundations for psychoanalysis that it would grow to conceal, only to be recovered later by theorists of media.

It is surprising that in the leap from organ projection to prosthesis, nowhere does the obvious connection between organ projection and the history of projection in psychoanalytic theory surface. This was a highly contentious issue for early members of Freud's psychoanalytic circle, because it once again touched upon the mechanistic physiology that was the forgotten history of psychoanalysis. The notion of projection allowed psychoanalysis to take seriously the powerful possibility that image technologies resolved the problem that plagued psychophysical research from the start, namely, how one could distinguish between the psychic apparatus and all of the other apparatuses that marked the machine-driven science of the late nineteenth century. In the hyper-materialistic framework of psychophysics, all machines were subject to the same rules of thermodynamics and all effects were localizable as specific operations of their internal mechanisms. This was obviously a problem for a theory with the conceit of making sense of the qualitative

and symbolic properties of human experience, which found no immediate correlate in the psychic machine. Projection remained a difficult topic
that Freud would revisit throughout his career, as it forced confrontation
with the relationship between the material nature of the psyche—which
was the province of science—and the symbolic operations that made it
a topic for interpretation.

As Freud wrote in 1923, with a distinctive echo of *Elements*, "The ego
is first and foremost a bodily ego: it is not merely a surface entity, but is
itself the projection of a surface," adding in a footnote that it "may thus
be regarded as a mental projection of the surface of the body . . . representing the superficies of the mental apparatus."[80] On its own, bodily
coherence is not enough for ego autonomy. The ego is only em-bodied
by means of a projection through which the mental apparatus and the
body as a surface get synced through their distinction from external reality. And the unity of the body as a surface is the result of having been
transformed into an image by an apparatus. This has three significant
and closely related implications with respect to Kapp's notion of organ
projection. The first is that projection is a technics that is at once material
and symbolic. The body cannot be recognized as real without a symbolic
register, and the ego as a symbolic unity is dependent on the materialities
of the body and psychic apparatus. Second, the conditions of unity are
image-based. The chaotic overabundance of the real gets sorted and organized into distinct "surfaces" or topologies that can be identified as things
like egos and objects by becoming two-dimensional images. For Kapp
the point of reference that remains hidden in Freud is explicitly the history of optical devices that gave rise to geometrical projection and linear
perspective. Optical instruments relied on the regularities of light and
lenses for producing images that then established standards for "realism" precisely through their divergence from the three-dimensional real.
Referring to the work of the painter and physiologist Carl Gustav Carus,
Kapp thus observes that "from the standpoint of organ projection" the
"*camera obscura*'s construction is entirely analogous with that of the eye"
in that it is an "unconsciously projected mechanical after-image of the
same, by means of which science is belatedly aided in comprehending
visual processes."[81] The body, in this case the eye, only becomes comprehensible through its externalization in an instrument, even as the eye
itself is the physiological apparatus through which images are resolved.

This underscores the fact that images per se are not enough to explain interiority, representation, and consciousness. Their explanatory power, for Kapp is in their association with an optical mechanism, whose material operations are consolidated and made sense of as a part of a single process. Finally, projection also resolved difficulties of ego formation by creating a boundary between the inner and outer world that was impossible to institute from a purely mechano-physiological perspective. Kapp argues that "there is no Self without body" and the "mechanical work-activity" is a "displacement outward of the inner world or representation" that encompasses the "totality of cultural means" whose re-entrance into inner life is a "self-affirmation of human nature."[82] In language that might strike a contemporary reader as directly lifted from psychoanalysis, he continues, writing:

> This self-awareness is effected through the human being's use of and comparative reflection on the works of his own hand. Through this authentic self-exhibition, the human being becomes conscious of the laws and processes that regulate his own unconscious life. For the mechanism, which is unconsciously formed on the model of an organic prototypal image [Vorbild], serves retroactively in its turn as the prototypal image through which the organism—to which the mechanism absolutely owes its existence—is later explained and understood.[83]

For Kapp, the boundary between the inner and outer world is established through a projective process. What some psychoanalysts call the "ideal Ego" assembles the machinations of the physiological apparatus into a psychological unity as an image. It is projected from the body outward, whereby, seeing a complete picture of the body's work in the world, all of the physical mechanisms required to bring matter to form undergo a backward projection—what Kittler referred to as "retrojection" and Sándor Ferenczi called "introjection."[84] Similarly, the projective function in Lacan's mirror stage, which places technological mediation at the center of ego formation, clarifies elements of organ projection that lacked a proper theoretical context at the time that Kapp wrote. The "formative effects" of the Gestalt made possible by the "mirror apparatus," which is at once a confrontation with the difference of an external image and a unification of the "I" with "the statue onto which man projects himself," frames the interplay of the real, the symbolic, and the technical in a way that is nearly identical to Kapp's understanding of organ projection.[85] By moving technics from the periphery of both cultural production and ego formation to their respective cores, the eternal conspiracy between the real and the symbolic that was so important in Lacan for later media-theoretical engagements with psychoanalysis appears at the beginning rather than the end of the story. In this sense, Kapp anticipated the Lacanian basis for cultural techniques which insisted on the inseparability of "operative chains" that "preceded the media concepts they generate," but from which they could not be divorced.[86]

THE LOGIC OF TECHNOLOGY

Kapp's philosophy of technology does not come into plain view until the final part of his book. It is only in the final chapters that he unfolds a theory that encompasses the autonomous machine and roots it in a relationship to the animal body. And it is only with this final twist in the theory of organ projection that its ambition to create an "evolutionary" theory of culture emerges.

Chapters 10 and 11—"Machine Technology" and "The Fundamental Morphological Law"—make up more than a third of the book. This is because they collectively treat a single but crucial form of organ projection, namely, the projection of the principle of animal organization itself. Kapp, we contend, is not arguing that animals and machines resemble one another and justify an analogy, nor is he arguing that they are potentially identical. He is arguing that the pure theory of machines makes clear an unconsciously known principle at the basis of the organism, a "measure" of the human animal. This turns "anthropometrism" on its head. It is not only that the human is the measure of things, but also that the technological process allows us first to take the measure of the human. These chapters rebound on the "anthropometrism" defended in chapter 1. In the unconscious interplay of machine production Kapp finds a new principle for the organism. This in turn allows the philosophy of technology to deepen not only the biological sciences but also the "logical sciences," as Hegel defines metaphysics. The result is a metaphysical vocabulary for the machine-mediated lifeworld.[87]

Kapp's Hegelian conception of machines was premised on Franz Reuleaux and his influential book *The Kinematics of Machinery: Outlines of a*

Theory of Machines (1875), in which he fundamentally reoriented the analysis of machines from dynamic to kinematic analysis.[88] Dynamic analysis treated the operations of machines as a deterministic series of specific forces that were calculable according to Newton's laws. In this view, a quantifiable force was applied to a part of the machine, which was then transmitted to another part of the machine, and then another, proceeding in a directional sequence. In short, dynamic analysis focused on the actual interactions between individual parts of a machine, whose collective function was the sum total of their forces on one another. By contrast, Reuleaux's kinematic analysis was visual, geometrical, and topological. Rather than concentrating on specific forces, he analyzed the abstract geometries of motion as delimited by closed sets of mechanisms. The basic unit of analysis was a "kinematic pair," which created a definite geometry of movement that could then be assembled into "kinematic chains" or a series of pairs that together produced new geometries of movement. A simple example of a kinematic pair is a screw and a nut, which together create a range of possible movements. The nut moves along the screw in a very specific manner that can be mapped and analyzed as a shape. From this perspective, the possibility for a mechanism's movement is made sense of as an image, irrespective of the individual forces applied.

Kapp, taking up Reuleaux's theory explicitly, writes that "the concept of the machine must be ascertained in order to be able to measure against it the representation that we make of ourselves."[89] We should not make too little of the word "concept" here; it is baldly Hegelian, and Kapp credits engineering with a truly dialectical achievement. The true concept of the machine, for Kapp, is an abstract notion. A true "machine science" is founded on Reuleaux's general definition of machinic operations, and these then determine the possibilities of embodied machines. Reuleaux is credited with a dialectical achievement, and Hegel serves to articulate the logic of the new technology.

More than a mere point of reference, Reuleaux's kinematic theory of machines acted as an epistemological foundation for Kapp's work. For Kapp, it demonstrates that "organ projection has a powerful ally in the machine," whose "knowledge-producing power . . . stabilizes and orients philosophical research" and "is intimately related with the body and the soul of the human being."[90] And as he continues in the same passage,

"Machinal kinematics is organic kinesis unconsciously transmitted to the mechanical; and learning to understand the original by means of its transmission becomes the conscious task of epistemology!"[91] Kinematics was nothing less than a skeleton key to the relationship between materialism and idealism, physiology and psychology, and the real and symbolic. It explained how discrete, mechanistic operations could be recognized as a unified form, shape, or geometry, which then made the collective interactions of those operations comprehensible. As a result, the unconscious was no longer an esoteric realm of spiritualism, but instead constituted the individual mechanisms prior to being understood as a kinematic pair—prior to being understood as an image. Reuleaux's terminology, including the relationship between the "technical frame" or "enframing [Gestell]" and "form [Gestalt]," which offered an alternative expression of the mutual dependency between kinematic analysis and kinematic pairs, would reach well into the twentieth century, most notably in the work of Martin Heidegger, whose notion of a "standing reserve" of energy caught in the metaphysical disclosure of an always "enframed" nature has played an influential role in the theory of technology since the middle of the twentieth century.

Kapp beat Heidegger to the metaphysical punch. The Gestell of Reuleaux's kinematics retroactively justified both a deepening of biological doctrine and a shift to a logic that could grasp both machines and animals—something we might call a "techno-logic," understood as the dynamic thinking that emerges from the notion of organ projection. "Machinal kinematics," Kapp holds, reveals to the retrospectively conscious mind what it has "transmitted." He calls this other process "organic kinesis," which is also the principle of spontaneity, the dynamic organizing principle at the basis of animal life—it is hard not to hear in this notion an anticipation of Humberto Maturana and Francisco Varela's influential notion of "autopoiesis," the organizational principle that dynamically retains its autonomy in the animal body, determining the components and allowing them to determine the organization in mutual self-production of the organism.[92] Kapp's principle of spontaneity, nearly a century before Maturana and Varela, allowed him to demonstrate the force of the theory of organ projection.

Kapp follows the distinction between machinal kinematics and organic kinesis with one between measure (Maß) and "metric" (Maßstab). Each is a manner of representing magnitude in a medium other than the one in which the measured object exists. The metric is expressed as number; the measure is "the reflection of a relation among orders of magnitude" in which "life processes are in motion, while metric is enforced from without."[93] Kapp returns to the hand to illustrate: "No organ comes with an arithmetic label attached. What is a finger on its own? A decomposing thing, soon to expire. But think of the many primitive stone axes that remain to us: each one is still the same as it was millennia ago. There is no finger without the hand. On the hand, the number five amounts to an indivisible whole, indivisible like the colors of a spectrum or the poles of a magnet."[94] The hand is not only the source of an imitation of its capacities but *also the ability to measure* that allows for science as well as the fabrication of artifacts. It is the source of discrete number *and* continuous proportion, of what would later be called the digital and the analog.[95]

Kapp argues that number can be construed in three separate ways: arithmetically (along the dividend or divisor line, in opposite tendency), in discrete divisible quantity (through the "harmonious" fractions), and geometrically (approaching zero and infinity, respectively). He derives "arithmetic" from Greek ἄρω, which he translates as *füge*, or "obey," "structure," "harmonize."[96] The suggestion is that arithmetic is not merely the rules for counting but the harmony of the balance between the different types of measure—a series of logics that emerge from the technological matrix and are cognitively coextensive with its progression. Kapp thus points to an expanded theory of number that includes proportion as quality in addition to quantity. To this end, he introduces a new pair of terms: *enumeration* (*Anzahl*, the German word for a counted quantity), "defined as the sequence of discrete units, adjoined consecutively with one another in a continuous arithmetic progression, *the unbound number*," and *innumeration* (*Inzahl*, not previously a German word), "defined as a *number bound* within the unit as difference"[97] Kapp integrates number into life by making numbers qualitative. Proportion, taken on its own terms, seems to offer a separate numeral system for biology. Projection of the whole body patterns the machine, the machine conditions our very knowledge of that body, and two separate systems of logic or even counting result.

It is tempting to leave it at that. Kapp has invented yet another dualism, this one for mathematics. But he frustrates the attempts—and there have been many—to reduce his thinking to organicism. Number is also not stable, not a source of thinking that remains untouched by the philosophy of technology. Numbers, Kapp tells us, "are not originally derived from external objects but instead emerge from the mysterious depth of the fundamental ratio in the body's organization . . . numbers must actually therefore originate in an inborn organic distinction. The representation of this distinction, projected outward from within, could then be transformed into the reality of numerability and the enumeration of individual things."[98] In other words, the very coming-into-quantity of thought and fabrication is itself a basic operation of organ projection. By steadfastly refusing to see material and ideal, discrete and continuous, organic and mechanical as *pre-technological* distinctions, Kapp gives himself the space to derive literal vocabularies from an original projective milieu. This makes the philosophy of technology Hegelian to the core.

Kapp never calls on Hegel to explain these distinctions, but it is plain that Hegel's *Logic* is the source code for this theoretical development. The *Logic*, we should remember, is not a formal but a material—or better, metaphysical—work. It details the development of the concept through a series of categorial developments under the headings "logic of" being *(Sein)*, essence *(Wesen)*, and concept *(Begriff)*. The transition from being to essence—the moment where the thing gets its identity—is caught in the development: quality, quantity, measure *(Maß)*. Quality is pure "in itself," a "breadth of existence, of something."[99] Quantity determines this purity, giving it boundaries in the form of magnitude. The "quantum" at the basis of this magnitude can be made discrete, in which case it becomes *Anzahl*.[100] Only *then* is the logical result *measure*, which Hegel defines as qualitative quantity. Some half-century later, we find Kapp using Hegel's vocabulary precisely. The *measure*, however, is itself split now, into *Maß* and *Maßstab*. In other words, if we trace the metaphysical stakes of Hegel's logic forward to Kapp's philosophy of technology, we find a "breadth of existence" now covered by machines. But Kapp refuses to talk of technicization, to see machines as fundamental. It is rather their operations or their *logic* that is foundational and evolutionary. Hegel had written that the understanding [*der Verstand*] and science operate within the reflected distinctions of essence but treat them as unreflected.[101] In this respect, we can see Kapp's work as forcing

technology into its original state of reflectedness, not as a linear imprint-
ing of human essence onto the world, but as the self-estranging externalization of the dynamic and unpredictable process of logic itself. If
objective spirit needed to include machines—and Reuleaux had demonstrated that beyond a doubt, for Kapp—then the logic of machines would
have to be integrated not only into the "culture" of absolute spirit but
the very logical operations of spirit itself. Machine and culture are one
operation.

CONCLUSION

The name Gilbert Simondon is today most distinctly associated with an
independent study of technology, a study of the elements, individuals,
and assemblages that machines take on in a kind of autonomous evolution. Simondon's influential synthesis of a study of technology imagines
a disciplinary discourse that would be like the study of literature, taking
its object as autonomous and irreducible to principles other than its own.
This study would aim at recovering a "regulatory control" over technology by stripping its discourse of passion, of the phobias and philias that
dominate the media's (and too often the humanities') accounts of technology even today. Simondon would have us live harmoniously with our
creations, rather than opposing culture to the technological.

This separation of technology from any analogy or source in the human
might seem very far from Kapp's "anthropometrism." But Simondon,
too, places technology at the origin of any world one might imagine:

> Man finds that he is bound to a universe that is experienced as a milieu.
> The emergence of the object can only happen through the isolation and
> fragmentation of the mediation between man and the world; and accord-
> ing to the principle proposed, this objectification of a mediation must have
> as correlative, in relation to the primitive neutral centre, the subjectifica-
> tion of mediation. The mediation between man and the world becomes
> objectified as a technical object, in the same way as it is subjectified as a
> religious mediator.[102]

Technology as semiotic milieu: Kapp embraced a position similar to this
posthuman vision of technology as more original than the worlds we
imagine it to inhabit. He simply added that the process through which

machines are generated, and the ineluctable technical condition of the human, was also logical at base, in the expanded Hegelian sense. If that logic was progressive, aleatory, proceeding according to patterns that could not be guessed in advance, then no fixed notion of "reason" nor any static physiological phenotype (not to speak of any political "real") could guide the human. The human was coeval with and subsequent to technology—posthuman in a genuine sense that somehow bridges poststructuralism and German idealism, engineering and political theory, Germany and Texas.

Late in his book, Kapp returns to the ax, using the following anecdote to demonstrate the consistency of his wide-ranging theory:

> I once looked on as an old backwoodsman in West Texas demonstrated what he called the *philosophy of the ax*. Placing his American ax beside one manufactured in Germany, he explained in rough terms the differences between them: the shaft of the German ax was straight as a post, rigid and refractory, while the slender "handle" of the American ax displayed a pleasing longitudinal undulation; the iron head of the German ax was attached at a rigid right angle, while the American ax head angled slightly inward and downward. He then dwelled on the latter with unusual appreciation, running his calloused hand over each individual swell produced in the iron as it was forged. He explained the ax's advantages, in terms of a powerful forward strike as well as an easy, relaxed backswing. . . . A tool like the American ax is perfectly finished to the extent that its formal relations are a perfect reflection of those of its organic prototypal image. This agreement affects the symmetry that imparts practicability to the object and lifts it, beyond the useful, to the sphere of the pleasing, therefore to the realm of the applied arts.[103]

What fell into Kapp's hands in his transatlantic adventure was not an imagistic analogy, not the simplistic notion that tools are extensions of function or imitations of form. It was the notion that the image *results* from technological activity, and that this image is the basis not only of the world of utility but also of the progressive and plastic knowledge associated with science and other forms of culture. The limited form of cognition that the idealists had called "the understanding" is here united with the higher forms of reason in a dynamic relation that is produced by and produces tool, artifact, machine—and ultimately language and the state. What *seems* an analogy is actually an embodied logical necessity, though not a predetermined one. The retroactive necessity of specific artifacts and systems is limited by the connection between the producer, the object of the force of technological necessity, and the logic of that process. All of these are united in "the human," who turns out to be extended, externalized, and posthuman after all.

ELEMENTS OF A
PHILOSOPHY OF TECHNOLOGY

Preface

The branch of technology referred to as "mechanical technology" is the central object of the present study. The recent tendency to treat empirical matter in a philosophical manner is encouraging evidence of the fact that empiricism and speculation have mutual need of one another. A philosophy of technology is therefore justified to the extent that the present reflections succeed in demonstrating that the emergence and increasing perfection of artifacts originating with the human hand are the primary condition for the development of human self-consciousness.

Because I do not feel equal to the task of providing a sufficiently comprehensive answer to the full scope of the problem, I have attempted at least to lay the foundation for it, should the following be accepted as such. Insofar as I believe I have found a new principle for incisive investigations into the problem, the present study proceeds from new points of departure.

First, incontrovertible facts will show that the human being unconsciously transmits the form, functional relations, and standard proportions of his bodily organization to the works of his hands and then it is only after the fact that he becomes conscious of these analogical relations with himself. This realization of mechanisms on the model of organic prototypal images [*Vorbilde*], the understanding of the organism that is achieved by means of mechanical apparatuses, and the exposition of the principle established as organ projection as the only possible means of achieving the objective of human activity—this is the actual content of the present work.

In what follows, we will return to the original meaning of the word *mechanical* and to its standard usage, limiting it to works of the human hand exclusively. In the process, we will encounter a conceptual confusion that, occasioned by exaggerations of the mechanistic worldview, alters the human being's correct conception of himself to the detriment of society. For the human being who truly believes in himself and his personality would, on the one hand, never confuse himself with a technical frame [*Gestell*] and, on the other, he would not tolerate the demand that he, the microcosm, be somehow degraded as a result of a disavowal of the difference between the macrocosm and, for instance, a pieced-together planetarium.

Moreover, nowhere is justification to be found for a mechanistic intuition of the things that are as intimately involved with the human being as are the tools he has himself manufactured and named in such a way that they serve him comparatively in elucidating organic relations.

As far as details are concerned, I would point out that the course of the investigation follows the body's organizational interconnectedness from the extremities to the internal organs and is rounded out in an account that leads us back to the beginning. In so doing, at no point will we transgress the boundaries of the domain within which the historical human being acts, and we will be careful to avoid digressing into regions where we cannot offer empirical evidence.

In order to provide the necessary technical and physiological materials, I have made as conscientious use as I could of the resources available to me, and I ask the reader's indulgence if, out of negligence, I have overlooked something essential.

In all circumstances, I had to put myself in a position such that I was able to perceive the gain in citing others' experiences and expressions, from which I was somehow able to infer a rationale for my own views. It will be readily apparent to the reader how much I owe especially to the works cited in chapters 6, 10, and 11.

In most instances, I have provided the page numbers for each citation. An exception to this is found in chapter 10; there an in-depth discussion of the book in question should be sufficient explanation for this.

Adhering strictly to the task at hand, I have maintained a proper distance from any polemicizing about the scientific questions of the day.

Only some of the illustrations accompanying the text did I create myself. I do not wish to neglect to express my enduring gratitude to the following publishers, on behalf of my own, for generously granting permission to borrow their illustrations—from R. L. C. Virchow's *Archiv für pathologische Anatomie und Physiologie* (G. Reimer Verlag), Franz Reuleaux's *Theoretical Kinematics* (Fr. Vieweg & Son Verlag, Braunschweig), Wittstein's writing on the *Golden Ratio* (Hahn'schen Hofbuchhandlung Verlag, Hannover), Küpper's *Apoxyomenos* (Collected scientific lectures for the common reader, ed. R. Virchow and Fr. Von Holtzendorff, no. 191, C. Habel Verlag, Berlin), Czermak's *Popular physiological lectures* (C. Czermak Verlag, Vienna), and from the works of Adolf Zeising (J. A. Barth Verlag, Leipzig).

One final remark: Today even books of moderate scope include a name and subject index, in addition to a table of contents. I have gladly followed their example, in order where possible to expressly facilitate a reader's orientation in a book as firmly internally coherent as this one— such that the book may be held jointly responsible for the breakdown in judgment of any individual.

Düsseldorf, January 1877

CHAPTER I

The Anthropological Scale

Though its object may vary as it dilates through space and time, thoughtful reflection never grows lonesome nor is lost to the infinite; rather, sooner or later it returns along the same path on which it embarked—that is, to the human being. In him, reflections link up without interruption, only to yield, after all seeking and discovery, the human being once more—the "thinker," in the most proper sense of the word.

It follows that the content of science, in the course of its research, would be nothing other than the human being returning to himself.

In this process, should human consciousness of the outside world enter into a relation of ceaseless comparison with the world within, then the human ascends to self-consciousness, knowing that his thinking his own being guarantees his difference from other beings.

What is meant today by the Self of which the human becomes conscious is not the same as it once was. The Self is no longer defined as the quintessence of a spiritual attitude alone. A fantastic illusion expires with the insight that the bodily organism is the Self's most immediate and authentic existence. If one were to disregard all the structures that together make up the living, human union of limbs and to think away its matter entirely, what else would remain of this vaunted Self but a spectral intellect?

Only with the certainty of its bodily existence does the Self truly enter into consciousness. It is because it thinks, and it thinks because it is. The word "Self," derived from the composite form *si-liba,* means "life and limb."[1] From this point forward, this basic meaning must be taken in complete seriousness. Not here in part, nor there in part, but the whole, undivided Self is at hand, in concrete self-awareness.

This manner of self-conception, unconsciously prepared in the passions and the senses and existing as a general mood, has reached the point where, through the indefatigable work of thought, its proper expression now emerges: one that fixes the new and makes of it a more or less conscious common good. The newest natural sciences in particular have been tasked with supplying evidence that, in the first instance, the bodily organism is responsible for the quality of all directions of human activity.

The natural sciences and philosophy—often by apparently antagonistic ways, though one may just as often be found in the borrowed armor of the other—each time they lose their bearings, find them again in the human being.

When the philosopher claims he knows no other world than that in relation to the human, the physiologist fully agrees, his vocation being to learn what lies within the human being and to corroborate the truth that all wisdom resides in knowledge of human nature.

Antiquity already made confident strides along this path. However, the prophetic foresight of poets and thinkers is one thing, the findings reached by doctors and scientists through focused, goal-directed activity another. What for antiquity had been the primarily unconscious beholding of a universal truth is for us today the conscious work of enlightenment and proof, through in-depth investigation of myriad detail.

Once exploration of the base materials of the world had occupied philosophy long enough, there came a premonition that the elementary natural phenomena were fundamentally attuned with human nature, and this precipitated the well-known words of Protagoras: that man is the measure of all things.

In want of physiological evidence, it was in antiquity less the bodily than the reflective being that was meant by this. Nevertheless, the *anthropological scale* had been definitively formulated and the true germ of human knowledge and ability identified, even if it remained deeply shrouded in obscurity.

Ancient Greek art owes to this insight the timelessness of its content, for its chisels gave form, in the semblance of deities, to the ideal human being. It is telling that sculpture, to which he devoted his youth, was for Socrates the precursor to the spiritual or ethical art he was later to develop on the basis of the famous inscription on the Temple of Apollo at Delphi: "Know thyself." Certainly, the whole of human civilization is, from its inception onward, nothing other than the stripping and disclosure of this germ.

The first attempts at explaining organic processes at this point took as their method the philosophical observation of nature.

Therefore we find Aristotle observing the body, concerned less with its external qualities than with how it acts as a means of manifesting the spirit within. The fact that his ancestors were doctors was crucial to his investigations and motivated his work on comparative anatomy and physiology. After Aristotle, physiological study and experiment was conducted by doctors almost exclusively, until very recently, when the outsize accumulation of materials required the work be divided among natural scientists and philosophers.

The history of this work is the history of physiology. If we understand knowledge of the bodily organism as knowledge of the Self, and if this self-knowledge and self-awareness is the source and ground of all other knowledge and ability, then herein lies more than a hint as to how we should treat the history of the discipline, for this discipline is appointed to provide and to go on providing all other disciplines with the material that is indispensable to their reform.

Great scientific discoveries are not associated by simple chance concurrence with epoch-defining historical events; on the contrary, they reveal themselves as the authentic driving force of such events. This is a genetic affinity in which the progressive disclosure of human *physis* is intimately involved.

Aristotle's description of organic organization and development as an "active generality" provides the first clear evidence of this inner correlation. Further evidence then follows from, among others, Galen's work on the nerves and brain, Paracelsus's concept of the macrocosm, William Harvey's doctrine *omne vivum ex ovo* and his discovery of blood circulation, Joseph Priestley's discovery of oxygen, Antoine Lavoisier's theory of respiration, Luigi Galvani's observations concerning the effect of electricity on the muscles and nerves, and, above all, from the astonishing successes of microscopic and chemical investigations into the nerves and the senses that are affiliated with today's most celebrated names.

Outstanding achievements like these should be marked, if we recall that the unity of natural forces reveals itself in the unity of all psychic functioning, as so many accretions to self-awareness and therefore considered essentially co-efficient with world-historical processes.

At this point, the philosophy of history has another great task awaiting it. In its execution, preliminary studies in "ethnopsychology" will be of equal value as methods applied to evaluation of historical fact from a physiological point of view, such as we find in the occasional article on the "physiology of the nation."

Psychology and physiology have been estranged long enough. The former may have advanced more rapidly, but the latter is quickly closing the distance, although even this is not yet enough, according to some: the one must absorb the other. What seems certain is that, if they are thought of as converging rather than isolated, united in a single course, they will flow into the broad riverbed of anthropology, there initiating a higher phase of self-consciousness, as "physiological psychology." This new discipline, in omitting nothing from its description of human being, will have the effect of stabilizing the fluctuating concept of personality, precisely by relocating the concept entirely within self-consciousness. The Self is the person; the self-conscious and the personal being are one and the same.[2] The modern cult of the animal, which originated in a misunderstanding of the theory of development, has of course criticized both, without at all considering that persons are the only ones capable of debating a concept of personality in the first place.

An earlier conception of personal being expressed it as the aggregate of two constitutive parts: a physiological part belonging to the natural sciences, and a psychological part belonging to philosophy. This conception is entirely in keeping with the dualistic classification of the cognitive process, which required that physiology and psychology proceed in parallel. Insight into the interconnectedness of the whole is, after all, only possible after treating each aspect of an object separately and securing in this way a fundamental understanding of the individual. The "two" that exclude one another in contradiction, in their difference include one another as "both." Thus the inalienable form of all cognition is and has always been based on a dualism.

Dualism both foments danger and generates the rescuing element; it helps to prohibit and to crucify, is crucified in turn and reduced to ashes; it is just as much the legendary Wandering Jew of the sciences as it is the mythic Proteus of thought. As pole and pole, as the matter and force that constitute the universe, en masse and in detail, dualism is "the spirit which eternally denies," "part of that force which would do ever evil, and does ever good."[3]

Dualism incites humanity in struggle and in need, severs church from state, and—divisive but also combinatory—affords reconciliation, progress, and pleasure. Just as the two-legged human makes headway only by alternating its steps, progress in general is possible through dualistic alternation alone. Of course, each opposing side thinks itself in the right. And while it is true that both are in the right, it is false that either alone is right. In fact, the more stubbornly each insists on its sole authority, the more completely will their content and the truth emerge—that, prior even to any consciousness of time, the two of them are one.

Centripetal and centrifugal tension—as well as their analogues: deduction and induction, idealism and realism, spiritualism and materialism—are dualistically operative throughout the entire gamut of world-historical conflict.

The golden age of German philosophy was eventually succeeded by recent achievements in the natural sciences. After the former's disdain for the latter had been reciprocated—with such hostility that it was said to be the "strangulation of philosophy"—today we are witnessing the natural sciences and philosophy join hands, uniting in order to transform and secure the foundations upon which, in perpetual struggle against the old syllabistic adversary of knowledge, the construction of a new, higher world order may be announced.

Without dualism, no scientific discussion will ever conclude, or, for that matter, begin—in short, without dualism there would be no scientific discussion at all. One simply cannot come to grips with the human being in a single go. He must be grasped discursively, pictured successively, first from one side, then the other. This method—of working first with intellectual qualities, next with the bodily—inevitably means we must accept a series of partial payments in a language of small change. That said, we trust implicitly, without need of express and repeated assurance, that science shall act prudently to safeguard the polar relations of opposites and their dialectical current against a one-sided rigidity.

This view should also prove useful in the work awaiting us here, for we have much to accomplish in a field where contradictions and misunderstandings are the order of the day.

To our mind, the anthropological scale lies, root and branch, in the whole human being. But, in scientific investigations and in language too, emphasis does tend to fall on either one of the two sides in which the human being appears. In the time prior to Kant, the psyche was decidedly privileged.

The standpoint held by Socrates and Protagoras is characteristic of the attitude antiquity had toward the psyche. The Middle Ages advanced a mystical manner of thinking, in which the intellectual intuition was the prevailing organ of cognition. According to Meister Eckhart, the father of German speculation, the person is the basic eternal form of all true being [Sein], and the human need only turn back toward his own individual being, his original nobility, in order to participate in the Most High: "If I knew myself as I should—that is, most closely—I would know all creatures. No one can know God who does not first know himself."[4] Adolf Lasson remarks on this: "The moderns may pride themselves on the presuppositionlessness of their pure thought, though they maintain the one presupposition indeed—the reality of the human soul, and this they share with the ancient Meister, who also accepts no other presupposition."[5]

Leibniz later writes, in a similar vein, that there will come a time when the great value of a sacred philosophy will again be recognized by *the human being turning back toward himself,* when the natural sciences will once again serve to glorify the author of nature, who presents to us in the visible world an image of the ideal.

As research into the visible world of the bodily organism advances, we are beginning to see that a physiological groundwork has been laid out for us in Kantian philosophy as well. Adolph Fick, for instance, is so persuaded of this that he has declared the Kantian standpoint in philosophy to be thoroughly physiological.

From here on out, whenever philosophy ascends to too dizzying a height, it will seek to recover solid ground and foundation by means of physiological supports. But that this path too has stretches that are not free of danger, indeed where there is risk of drowning, has been proven for us by the doctrine (masquerading as philosophy) that truth—the only truth—should be sought in sensuous reality alone.

And so, for Ludwig Feuerbach and for others, the human being was *the* point from which all knowledge not only proceeds but to which all knowledge amounts—though not the human in a general sense, only the bodily human being. Hence the precept in his philosophy: man is he "who generates thoughts from matter, from existence, from the senses, a real being, or rather a being of the highest reality, the true *ens realissimum*," in the sense of a radical material coarseness, a human being of flesh and blood, a being who is only what he eats.[6]

Previous to this, the material side of our maxim that man is the measure of all things had been sorely neglected—so much so that the overhasty demand for positive corroboration on the part of the natural sciences far overshot the line of equilibrium. The natural sciences and philosophy now each had to yield somewhat to the other. The former finally achieved full consciousness by way of philosophical reflection on the meaning and coherence of its tremendous findings; the latter drew from the wellspring of empiricism a profusion of fresh and convincing evidentiary force.

The recurrent demand that the human being's point of orientation vis-à-vis the world be the human being himself could only ever be partially satisfied, so long as the notion of the human was restricted to the side of spirit alone. This notion required a solid physiological foundation in order for the demand to be fully realized.

Among the advocates of a physiological foundation for the human being—whose ranks are increasing daily—Oscar Peschel remarks in a lecture on the scientific merits of craniometry: "Nor should we be surprised that human research turns so late to the study of human beings, since it is only very lately that we are capable of taking up this last and greatest task." Speaking to geological questions, Edgar Quinet agrees: "In a certain respect, it would have been the more philosophical path, had we taken ourselves as the starting point—that is, a point with which we are most intimately acquainted."[7]

Constantin Frantz also remarks, with reference to the findings of Auguste Comte's *Positive Philosophy*: "Of course it finally occurs to Comte that the human being himself is the noblest object of research, and here he begins with comparative anatomy and physiology; for matter is the substratum of all appearance and, for Comte, thinking amounts to brain activity. *Beside matters of anatomical and physiological fact stand only matters of historical fact*."[8]

It is in our best interest to cite others who also speak precisely to this point, in order to demonstrate its importance from the broadest possible range of perspectives.

Karl Ernst von Baer writes: "The human being cannot measure things but by proceeding from himself and taking himself as the norm. In this way, the human being has learned to gauge space and also time." In the periodical *Ausland*, it is reported that with the use of physiological apparatuses, one is able to descend into the interior depths of the living body, indeed to behold there one's own self. Jean Paul says that consciousness of our selves is the key to the world. Lazarus Geiger states: "That which time and again most vividly captivates, most warmly gratifies humanity is the human being." Heinrich Böhmer calls physiology "the teleological science *par excellence*." And Adolph Fick declares one's own consciousness "the sole proper and sole possible point of departure for philosophizing."[9]

Adolf Bastian also speaks up for this beholding of the self: "The alleged opposition between spirit and body disappears in their harmonious union. Today's sciences are flourishing with life, because the spirit is rooted in the natural soil of the body. Knowledge that has been nourished at nature's breast is unfurling its buds along every branch of research, and the trunk of scientific inquiry extends its protective shade even to the fruit of self-knowledge that promises to ripen in the breast of man." In all his reporting, the notable ethnographer never allows us to forget that, even when taking the ideal view of a nation's cultural conditions and their aims, he always has one foot on the ground, on the earthly soil, and that, for him, real idealism and genuine self-knowledge are inextricably linked.[10]

In light of the agreement among these perspectives, we can see well enough how indelibly a certain credibility has accrued to the concept of the anthropological scale. The concept was not readily available, but instead had to be realized by way of an ever expanding detour through collection and research. The human who in thinking proceeds outward from himself is the requisite condition for the human being returning to himself. Hence thinking, like breathing, is an uninterrupted process of taking in and issuing forth.

Certainly we can distinguish prolonged periods of time during which one or the other tendency alternately predominates. But, in the highest and final instance, what brings reflection home is always the human being, scattered as he is over the face of the earth; and what physiologists and psychologists are gradually elaborating is nature encapsulated in humankind. The human being takes from the whole of nature to gather himself together; on nature's ground, the human being philosophizes his way to self-consciousness. The outside world gives the human being purchase on the world within himself. The content of one is the testing for the content of the other.

The discovery, reserved for our own day, of the *unity of the fundamental forces of nature* goes hand in hand with the disclosure of the *unity of human nature*. To the extent that the human becomes conscious that the unity of his own being has always been the unconscious foundation of his research into the coherence of the fundamental forces of nature, and to the extent that the human being thinks himself in and of nature rather than above or outside of it—to this extent, his thought is the coincidence of his own physiological disposition with cosmic conditions.

The center from which, centrifugally, thought emanates and to which, centripetally, it returns is ultimately situated within the human being.

In context of the universal scientific importance of the anthropological scale, the so-called *anthropocentric standpoint*—according to which humanity sees itself at the center of the world—can hardly appear senseless or without merit.

The human being exclusively—indeed absolutely—occupies the center of its own sphere of thought, whether this belongs to the individual or to the species as a whole. Its sphere of thought is its intellectual world, is and remains its world. For the human being, there can be no other world than that which exists in his representation of the world. In this cosmically extended egoism, the human being upholds the singularity of his species and its belief in itself. The representation of an infinitely expanding universe is possible only—if at all—on the basis of an I that represents and understands itself at the center.

Inarguably, the human being has more right to exercise this priority for himself than for another, alien creature, allegedly of a higher and more perfect nature—existing heaven only knows where. The human being does not actually nor could possibly know anything about such a creature apart from what he knows already with respect to himself. For the law of nature stipulates that, whether by sheer fantasy or the most advanced learning, no being is capable of surmounting itself, of breaking through the boundaries of its power of representation, a power permanently held in anthropomorphic and anthropopathic thrall. Anything lying outside its power of representation is unavailable to the human being and cannot be taken into account.

To represent and to think are anthropocentric behaviors in and of themselves. Every I is the center of a world. Anyone opposing the anthropocentric standpoint surely still stands on some ground of thought! Are its opponents not therefore unwittingly contesting the very grounds on which they stand?

Consider, in this context, the *geocentric standpoint*, long since overcome in astronomy: next in relation to the solar system familiar to us, it is self-evident that the sun, and not a planet like the earth, occupies the center. Apropos the limitless universe—in which all radii are of the same length and every point is a center, such that ultimately there is none—even the most foolhardy research stops short of positively identifying an astronomical center, content instead with maintaining a hypothesis that would shift the center of gravity of the system of fixed stars into the Pleiades.

But so long as the human being persists as the sole certainty for humankind, this self-certainty is then the only conceivable point to which the human being is able to affix the lever of his thought.

This no human being may renounce without also renouncing himself. This point of self-certainty is eternal and ever the same, supporting the human being in his construction of the arts and sciences, without which, however, the human being is like vermin, degenerate and debased.

There remains a trace of the geocentric in the anthropocentric point of view, in the following sense: after all the supports of its former sidereal priority have crumbled, the earth still appears as if tacitly taking part in some ideal centrality, for which the human beings inhabiting it are liable. Sensible appearance continues to exercise a silent, enduring power—the sun rises, the sun sets; this is not yet overcome, neither in representation nor in everyday speech.

The geocentric view has for centuries influenced the progress of culture. The ancient Greek—steeped in the conviction that his earth was situated at the center of the world and he himself at the center of the earth—created an intellectual world corresponding with this sense of himself. His resolute confidence in sensible appearance and in Rome itself as the center of culture, whether under imperial or papal rule, also had an unmistakable effect on the spread of civilization and the establishment of global dominance. As Paul de Lagarde trenchantly remarks: "The entire ecclesiastic mythology is rendered obsolete once the earth is removed from the center of the universe and becomes a mere mid-sized planet orbiting a minor sun. When the church first heard the words *e pur si muove*, the entire orthodox system was affected."[11]

The concept of centrality differs, however, when referred unilaterally either to the earth or to the human being. With respect to the earth, one of course always meant the real, spatial center. With respect to man, on the other hand, the word acquires a strongly figurative meaning, as, for instance, one might describe a leading personality as the soul or the center of a society. The real sense of the word, then, would be the locus—to the extent that this is assumed to be the highest and most preeminent, as that which determines all else, as the attractive core, as the spiritual focal point.

That is why the anthropocentric standpoint recognizes the human being as the culmination of an entire developmental sequence of organic life-forms, as the crowning achievement of creation. As soon as the earth ceases to be simply a planet for us and instead is regarded more broadly as the bearer of human spirit, and the moment this spirit appears in its unity—the union of mind with the individual body, and the union of humankind with its planetary dwelling—then the above-mentioned disparity in the concept of centrality escapes any stricter scrutiny. What is clear, though, is just how closely related the anthropocentric standpoint is to the one that regards the human being as the purpose and aim of planetary development.

The perennial opponents of the so-called teleological worldview continue to resist, fortified by a materialism newly supplied by Darwin. Whatever form this latest phase of the timeworn conflict may assume, both camps presently agree that the human being represents the apex of organic development.

Though diametrically different in approach, this is where the two paths converge—the one by means of mechanism and chance emergence in causal succession, the other by means of spirit and purposive creation. Nevertheless, they part ways again on the question of permanence. The one cannot grant the human being the position of apex for all time but keeps open the prospect of an even higher stage of development; the other believes, in the words of Carl von Rokitansky, that "with man, our lot is finished."[12]

An apex, as the superlative point and highest stage beyond which it is thought impossible to proceed, is at the same time the actual apex as well as that point in the series that in itself realizes every stage through which it has already passed—every stage whose truth it therefore is.

For the apex is not merely another stage gradually differentiated from those that came before; it is something qualitatively different. It is not a stage from which another, higher one is expected to emerge, but is the one and only existing ultimate stage for the sake of which all other stages have come to be. Because the apex is not a stage in the same sense as are the stages preceding it, and because in it the concept of stages must terminate absolutely, the conclusion of organic development cannot be an animal; its conclusion is the human being.

The human being is not a rung on the ladder of animal development but, as the last rung that does not lead to another, it is, as we have said, much more: in ceasing to be a rung, the human ceases to be an animal at all. The human being is the goal attained, the goal immanent in all prior stages—the ideal animal, as it were. No apex is thinkable without a base, without a substructure there is no apex; there is no human without the animal, no animal without the human. Only one assumption is left to us: that the idea of the human is the matrix and primary basis of every living thing.

The conflict between the two above-mentioned trajectories in the theory of organic development will not be put to rest before it has penetrated and fertilized every region of human thought and activity. The Lamarckian theory of descent, revived and adapted by Darwin, is *the* scientific issue of the day. Its proponents include natural scientists, its opponents philosophers, following the rigid opposition of nature and spirit. Its relevance is felt in all spheres in which both powerful camps operate. Religion, art, law, ethics—all appear shaken to the core by this issue. Indeed, even in the field of linguistics, Lazarus Geiger—a prominent figure and author of a work on the *Origin and Development of Human Language and Reason*—attests to Darwin's priority; in his book *Physics and Politics*, Walter Bagehot applies the principles of natural selection and inheritance to political society; and Carl du Prel, in his *Struggle for Existence,* applies Darwinian principles to cosmic processes.[13]

Without coming any nearer the conflict ourselves, we'll make do by positing on more neutral ground a *biogenetic law,* first articulated in 1812 by Johann Friedrich Meckel. The law states that the principal morphological transformations established in the theory of organic development are perceptible in the progressive stages of human embryonic being.

Ernst Haeckel summarizes this basic law as follows: "*Embryonic history* is a compendium of *phylogenetic history*; or, in other words: *ontogeny* is a brief recapitulation of *phylogeny*; or, somewhat more explicitly: the series of forms through which the individual organism passes in the course of its progress from the egg-cell to its fully developed state is a brief compressed reproduction of the long series of forms through which the animal ancestors of the organism (or the ancestral form of its species) have passed, from the earliest periods of so-called organic creation to the present day."[14]

The animal species therefore represent the stages of collective organic life generally, however indeterminate may be the intervals of time and space separating their emergence. By way of comparison, prior to birth the human being contains all preceding stages of organic development crowded into a minimum of time and space, though only the principal stages are recognizable in the embryonic growth of the individual. As Gustav Jäger writes in *Wonders of the Invisible World,* "nature has laid away in the developmental process of each individual a copy of every fully developed form that ever inhabited the earth, and to this day nature exhibits these to us as transitional embryonic states."[15]

In light of this idea, Lamarckian-Darwinian theory may achieve an understanding that is not openly hostile to spirit and its aims—for it asserts both purpose and development. We will not concern ourselves for the time being with the question as to whether this theory is justified in simultaneously assuming a fatalistically random configuration of matter, where there can be only incidents but never purposes, events but not development. We accept purpose and development and furthermore conclude: because the embryonic developmental forms are integrated into the overall development of the human being, in his life in the world, the human being therefore, *omnia sua secum portans,* everywhere and at all times carries within himself every antecedent stage of animal development as well, even if only in outline and remarkably compressed in space and time. This realization comes to us all the later for having been veiled, on the one hand, in macrocosmic obscurity, and on the other, in microcosmic obscurity.

Just as the mature human being exists in the embryonic cell as an original disposition, so too the idea of humankind is the germ or the original disposition of all organic creation. As always, the aim is also the primordial beginning of development, and the primordial beginning is likewise development's purpose and aim—development itself being the unity of a continuous progress, both forward and back. Thus Aristotle's ancient truth: "That which is later according to emergence is earlier according to idea and substance."

The human therefore stands in precisely the same connection with the animal world as he does with the embryonic stage of his own being, and the animal kingdom is the same past life for humanity *en masse* as the embryonic field is for the individual. The mature human is no more uniquely isolated from the embryo than the animal is from the human. Could it be, then, that the human being was somehow already present here before he was ever "born into the world"? Just as the embryo is the primordial world of the living, breathing human being, the primordial world of humankind is found in animal creation. Upward through the stages of animal life, in increasingly complex metamorphoses of matter, the spirit advances to the apex, at which point, apperceiving itself as self-consciousness, it knows it is also animal and, in this knowledge, ceases to be animal. A passage from Pascal celebrates how it is that such knowledge of natural necessity actually lifts the human being above it: "Even if the whole universe take up arms to destroy him, man—the weakest in nature—would still be nobler than his slayer, because he knows that he must die."[16]

It seems opportune at this point to revisit a relevant passage from my own *Philosophical Geography*:[17]

The human being is the general culmination of an entire vast sequence of stages of inorganic and organic creation—a sequence in which the first, least perfect form is absolutely as crucial as the last form approaching the human. Without connection or relation with his natural antecedents, the human being would indeed be a kind of physical abstraction!

In fact, the human being is in the most multifaceted association with the whole of nature and, having thus grown through kinship with it, is its concrete microcosm.

One may very well ask what befits the human being, insofar as he is an altogether natural organism, though not what would befit him to the extent that he is like a fish or a bird. The human is not born *of* but *after* the animal. In terms of development, chronological sequence may place him nearer a particular species of animal, but his inner relation with the whole is an indissoluble one that permits of no division. The earth is often referred to as the mother of humankind, but we see how, in reality, the idea of humankind is the mother of everything on earth.

The human being is therefore the regulative principle in nature. Nature should not stand above the human being, as it did for the ancients, nor should the human being think too highly of himself, think that he stands above nature, as he did the Middle Ages. Rather, the human being should be aware that he is both the substance and truth of nature, that he cannot accomplish his task by treating it with contempt or hostility, but only by striving to infiltrate and to understand it.

To the same extent that the human being is capable of comprehending the phenomena of natural life, his consciousness of his own nature will increase; in this way, he will realize that the development taking place "inside of us" follows the same laws that govern everything taking place "outside of us." Language, on behalf of self-awareness, has fit together in the expression "the nature of spirit" those two aspects that are all too often portrayed in dualistic contradiction.

The human being responds to the question "What is man?" No animal responds to the question "What is animal?"

At this point, the objective, gradual differentiation of organic nature terminates and the human being begins the subjective, essential differentiation of himself, as the apex and closure of the series of stages that led to him.

In order to avoid getting entangled in recent controversies, we would like at this point to concede in advance all of the approximately human traits that have been provisionally attributed to the animal: collective consciousness, language, a sense of justice, the formation of general concepts, a disposition to music and architecture, indeed morality and social improvement. For the moment, we remain undecided as to these assets—as well as to whatever else modern animal psychology should like to add to its inventory—in order keep clear of any anthropopathic confusion and to take up our inquiry into the limit that separates, once and for all, the animal from the human—the limit of *self-consciousness*.

However the findings of investigations into consciousness and its step-wise development may diverge, we find perfect agreement on at least one point—that the unconscious, the conscious, and the self-conscious in human nature are linked, one with another, in a genetic sequence of transitions from dark to light, and in such a way that the earlier is pre-served within the higher and remains operative in it.

The conscious—in relation to which the unconscious and the self-conscious act as the pre- and post-conscious, and which is understood as an individual's knowledge of a world external and imparted to him through his sense perception—is initially the relation between a perceiv-ing subject and a perceived object.

In the state prior to the awakening of consciousness, there is no dis-tinction at all between an inwardly perceiving being and an outwardly perceived thing. By contrast, at the self-conscious stage, the perceived object is situated within the subject, is itself something interior, some-thing constituting the I. This I, without further immediate communica-tion from the senses, is itself the perceiving and perceived thing.

External things enter into the human being as objects of his con-sciousness. To the extent that he discovers himself elucidated in them, they become his interiority. In this way, they deliver the I both from the dreams of the unconscious and from the spell of a dualism that opposes an interior to an exterior. In his interiority, knowing and known, subject and object are one, as self-consciousness.

Self-consciousness proves to be the result of a process in which knowl-edge of an exterior is transformed into knowledge of an interior.

This knowledge, turning back toward the exterior and expanding our understanding of it, in turn provides new information about our interior, ultimately producing, in this endless complication of our orientation in the world and of our self-orientation generally, the content of all knowl-edge [*Wissen*]—in short, the sciences [*Wissenschaften*].

Above all, one should be absolutely clear about what is meant by the concept of an "outside world." It is not as easy as is often presumed, not even for sense perception, to clearly demarcate what is "inside of us" from what is "outside of us." We are dealing here with a contentious border zone. Depending on given relations, the I dictates what counts as outside.

On certain occasions, the I existing in a particular bodily organism, or existing simply *as* organism, may claim all the constitutent parts of its body as its inner world. On other occasions, it may define the hand and the foot as belonging to the outside world, in the form of "extremities" that can be sensibly perceived like any other natural thing, a stone or a plant or the like.

And yet, without doubt, the entire body belongs to the inner world. Even if we regard the brain as the exclusive seat of thought, as the intellectual interior, we cannot therefore disregard the heart and the spinal cord, for instance, for the brain is incapable of thinking for itself; the *entire organism* unconsciously thinks with it.

There is another distinction concerning the concept of an "outside world" that is important for us at the moment, for we also have to deal with objects that the term "nature," as it is customarily used, is inadequate to describe.

The outside world comprises myriad things that, though nature did supply the material, are less the work of nature than the work of the human hand. As artifacts, by contrast with natural products, these make up the content of the world of culture. The "outside world" for the human being is therefore composed of both natural and man-made objects.

Animals too have an immediate sensible perception of things. But the animal does not comprehend what it sees and hears, smells and eats; these things remain something other and alien to it, an antithesis in which the animal persists.

But the human being overcomes this antithesis precisely because his original disposition affords him the competence to productively and receptively expand *ad infinitum* the sensory ability he shares with the animal by means of mechanical supports—in other words, through the works of his own hand. The human being is given to dealing with things, making use of them, forming matter in accordance with his advantage and subjective need. In so doing, the conscious and the unconscious are both equally operative—the former in his specific intention to remedy a momentary deficiency, the latter in specifying the form this remedy will take, albeit without a clear representation or volition.

The first crude tools were adapted to enhance the power and dexterity of the human hand in separating and combining matter. The modern human being lives within a highly cultivated and diversified "system of needs," such as we find summarily expressed at the World Exposition, for instance.[18] The human being sees and acknowledges in all of these external things—as he does not in the unmodified products of nature—the structure of the human hand, the deeds of the human spirit, both the unconsciously finding and the consciously inventing human being; in short, himself.

This takes place in a twofold manner. On the one hand, every tool—understood in the broadest sense of the word as a means of enhancing sensory activity—presents the unique possibility of moving beyond the immediate, superficial perception of things. On the other hand, as the combined product of intellectual and manual activity, the tool is so fundamentally and intimately affiliated with the human being that he finds himself beholding something of his own being in the creation of his hand, his world of representation embodied in matter, a mirror- or after-image [Nachbild] of his interior, a part of himself.

But since there is no Self without body, since the Self is both life and limb, as we saw earlier, this outer world of mechanical work-activity that proceeds from the human being must be conceived as the true continuation of the organism and as the displacement outward of the inner world of representation.

A depiction like this, of an outside world encompassing the totality of cultural means, is effectively a self-affirmation of human nature. But when the likeness withdraws from the exterior into the interior, self-affirmation becomes self-awareness.

This self-awareness is effected through the human being's use of and comparative reflection on the works of his own hand. Through this authentic self-exhibition, the human being becomes conscious of the laws and processes that regulate his own unconscious life. For the mechanism, which is unconsciously formed on the model of an organic prototypal image [Vorbild], serves retroactively in its turn as the prototypal image through which the organism—to which the mechanism absolutely owes its existence—is later explained and understood.

Only by way of this detour—through its spontaneous creation of cultural means—does humankind deliver itself from the base consciousness of sensation and attain the higher consciousness of thought and self. Accordingly, we differentiate between the tool's external purpose and the internal conception of its manufacture. The former is present to consciousness, the latter takes place unconsciously; intention guides the former, instinctive doing the latter. Both aspects, however, meet in fitness of purpose. This happens in the act of measuring: the limbs of the organism are engaged in realizing a purpose, and at the same time they provide the measure of a tool's practicability.

The testing of tools for potential inadequacies and the endeavor to improve them leads first to a comparison of the tool's purpose with the dimensions and proportions given in the body; next to the discovery that the conception of the tool that has yet to be made is already unconsciously aligned with the rule of functional relationships that governs the bodily organism; and finally to the certainty that all cultural means, no matter how crude or refined their construction, are absolutely nothing other than organ projections.

The craftsman's tool, the instruments of art, the apparatuses employed by the sciences to measure and weigh the smallest particles and velocities, even the sound waves formed and set in motion by the human voice: all logically belong to the category of projections formed of matter that I believe to have properly designated as *organ projections*—regardless of whether greater emphasis is placed on *physis* or *psyche* or both from a monistic point of view.

A *review* of the foregoing now yields the following findings:

We began with the concept of self-awareness, at which one arrives only through a full appreciation of corporeality, and with the human being's right to appropriate himself the world over as the exclusive measure of things—on which basis, in Julius Bergmann's apt expression, "the Not-I is, as it were, ideally appropriated by the I from the outset."[19] We then pursued the relation between physiology and psychology to its psychophysical apex. We considered the anthropocentric standpoint in its passage from sensible appearance to the ideal realm of the teleological worldview—a worldview that measures the start and aim of cosmic evolution against the self-conscious human being. Following this, we examined self-consciousness, on the basis of its distinction from the conscious and the unconscious, as well as the world of external objects that gives rise to it, to which domain all free mechanical forms nominally belong. We then drew conclusions regarding the human hand, from which all tools and implements proceed, and regarding the bodily organism generally, which, while forming itself, is also constantly producing and projecting itself, according to the following axiom: from each emerges only what already lies within it!

We will now attempt, in the explanation that follows, to establish a new, actual point of departure for epistemology, intended to make appreciable the truth that the world of culture, as it refers back to the projecting Self, is uniquely appointed to shed light further into the darkness in which the most vital processes of organic life still remain shrouded. The human being recognizes, in the second instance of measuring himself against things, that he is, in the first instance, the measure of all things!

CHAPTER 2

Organ Projection

In the early 1860s, at a meeting of the Berlin Philosophical Society during which the age of the human race was being discussed, Carl Schultzenstein remarked that everywhere the human being has appeared, he has been obliged first to invent a suitable way of life for himself and to procure this by artifice, in such manner that, in the human being, art and science take the place of the animal instinct. In this way, the human being becomes the creator of himself, even of his body's formation and refinement. To which Ferdinand Lasalle assented, replying: "This *absolute self-production* is precisely the most profound point in humankind."[1]

Here we have happened on a remark that will be useful in articulating a concept of projection in the sense that we would like to understand it.

Use of the term "projection" adheres strictly in all cases to its basic etymological meaning. In the military, missiles are called projectiles; in building design, one speaks of architectural projections; and apart from projects undertaken in the world of business, the term is especially endemic to draftsmanship, referring to every sort of sketch, map, plan, blueprint, and in particular to the cartographer's grid. Who, for instance, is unfamiliar with the parallel- and meridian-lined grid known as "Mercator's projection"?

Beyond these specific instances, what is of greater interest is just how frequently physiologists and psychologists use the word to describe the relation of the sensations to external objects and to describe the formation of representations [*Vorstellungen*] in general.

In all of these cases, *to project* is more or less to throw out or forth, to place before or displace outward, to relocate something interior into the outside world. When taken literally, projection and representation are not actually all that different, insofar as the inward act of representing is not independent of the object placed before the eyes of a representing subject.

A few instances will suffice to demonstrate how the expression thus far has been used in scientific works. In *The Independent Value of Knowledge,* Carl von Rokitansky writes, with reference to the displacement of things in space: "We are never conscious of the images inside of us, but only of the things we have projected outside ourselves." Carl Gustav Carus is a bit less direct: "In seeing, it is the object's own action of light that we perceive internally, and not an *image* of this action of light that another would also see. The visible world emerges for us only as the sensation that is aroused in the retina projects itself outward, as it were." Commenting on Ludwig Feuerbach's anthropological standpoint, Karl Rosenkranz alleges that the human being projects his own nature in the form of a representation of a subject existing outside of himself, which should be differentiated from him; and that, when it comes to religion, the human being estranges himself from his own nature by projecting representations that correspond to no reality.[2]

In his *Elements of Psychology,* Frederik Anton von Hartsen briefly describes the projection that is thought to be the soul apparently stepping forth from the body as the *"expulsion* of spiritual attributes." He writes about the projection of both sensations and desires and claims he is able to prove that the projection of psychic attributes takes place not only in space but in time as well.[3]

So far, the theory of projection has played a significant role in the study of spatial representation and the direction of vision as well as in explaining the upright position of the perceived object with respect to the inverted retinal image.

Generally speaking, depending on whether projection is accepted or rejected, two theoretical camps emerge, which Helmholtz has identified as "empirical" and as "nativist" theories. To the former belong, beside the Herbartians, Helmholtz and the majority of physiologists, including Johannes Müller, Carl Ludwig, Otto Funke, Carl Lange, and above all Friedrich Ueberweg, whose treatise "Toward a Theory of Vision" effectively revitalized the subject after a period of apparent dormancy. Even more recently, the subject was taken up in Eduard Johnson's contributions to the *Philosophische Monatshefte*, in Paul Kramer's "Notes toward a Theory of Spatial Depth Perception," and in Carl Stumpf's book *On the Psychological Origins of Representations of Space*.[4]

But the dispute remains unsettled. Ludwig has significantly suggested that "the verb 'to place outside' is only a figurative expression designating a phenomenon whereby the psyche *correlates* a condition present in the brain to an object located outside of the eyes, as its cause."[5]

One such correlation between an image [*Bild*] in the sensorium and an object outside of it is that of fact. This is, to say the least, most closely affiliated with the projection. Though he does not use the word itself, even Johannes Müller seems with a "quasi" to want to avoid abandoning the concept entirely: "We can conceive of the visual representation as a quasi displacement forward of the entire retinal field of vision." Wilhelm Wundt proposes a slight restriction to this view, "which presupposes an inherent or at least a strictly given correlation of the retinal points with the corresponding points in external space." Adolf Horwicz also clearly designates the projection "as the outward displacement of sensations into the object, as the correlation of the same with an external object."[6]

Having suggested that there exist absolute correlations of this kind—in which an internally necessary relation inheres in both sides, each of which is constantly confronting, presupposing, and projecting the other—we will leave the issue unresolved for the time being, and we will turn instead to a process that truly warrants being called projection, because all that underlies it are the sort of facts that preclude all difference of opinion. A detailed elaboration of this type of projection—*organ projection*—will justify itself in the course of our investigation, the actual theme of which it is.

The underlying facts are familiar, historical, as old as humankind. But what is new is our treatment of them, proceeding from their genetic context and from the point of view of projection that is here applied to them for the first time.

This previously untrodden path leads straight to the historical and cultural foundations of epistemology in general. Our point of departure is the human being, who, in all he thinks and does, unless he breaks with himself entirely, can proceed from nothing other than his thinking, acting self. We are not dealing with a hypothetical *bathybius-being* nor with a hypothetical ideal human being, but with the human to whose being may attest only the traces of and changes in the things he has made with his own hands.[7] This being alone is the fixed point at which all knowledge begins and aims. At all times, in all places, he attests to himself!

The boundary that had been assumed until very recently between a historical and a non-historical age, determined to the year on the basis of the biblical record—this has been set fluctuating, from the beginning of terrestrial creation, in centuries-long blocks of time.

Cave findings tell a story that is no less irrefutable than rolls of papyrus and libraries of clay tablets. They constitute a very real literature, a lapidary and pictographic script consisting of fossils and pottery, tools and rudimentary markings. This testimony lets us infer the nature and condition of human and animal kinds, as, under primitive conditions of competition, they fought simultaneously to wrest life from one another and to establish dominion.

In light of these findings, and of others that are unfolding new labyrinthine paths for contemporary linguistic research, the concepts of the historic and formerly so-called prehistoric are blurring to the point of indistinction. Those who pursue a distinction or demarcation of one sort or another are going to have to get used to the idea that the actual prehistoric human is a being no trace of which exists, not even in the crudest tool—for with the first tool history commences, because it is the first work. Insofar as history is understood as the progression of human work, the first work is, to say the least, the inception of something like history. Hence, incipient history only begins to take recognizable shape as history proper where a separation of workers by vocation begins to appear in the division of labor generally, in preparation of the eventual fixed division of the members of the social body into castes and the state body into estates.

All work is activity, but only conscious activity is work. Animals do not work. In the so-called animal colonies of ants and bees, one finds a division of mere industriousness. The division of labor, of conscious vocational work—that is what comprises the historical state and is already history.

Between actual prehistory—that is, human being prior to all history—and genuine history, incipient history has been assigned its place.

In the works published on this subject from the standpoint of a theory of descent, more often than not an author will tend to supplement with outright fantasy his depiction of the physique and the way of life of primitive human beings. As a result, one is asked to imagine the primitive human being at one moment as a brute among brutes, at another as an animal-like creature endowed from the outset with the germ of historical potential.

In view of the rude and savage animal world, in the vicinity of which we have to imagine the "budding" human being, we cannot overestimate its physical characteristics. Doubtless he displayed gorilla-like strength and agility. He must have possessed the enormous strength of storied historical men or modern-day acrobats, though of course he would have lacked their artistry and skill. Our own modern athletes' occasional feats of strength must have been, for primitive human beings, an ordinary, unremarkable, natural aptitude shared by all.

So long as the human being had to confront rapacious beasts unarmed, he needed to be able to match them in the strength of his bite and nails, in the power of his fist and arm, in the simian speed with which he moved. Imagine the force and dexterity needed to beat a steer to the ground using only a fist, to break iron with one's hands, to hold a hundred-weight from one's teeth, to swing from a trapeze and walk the tightrope. Imagine a single human being possessing all of these powers and you will have an idea of how the primitive human was physically fitted to survive his quite literal life-or-death struggle with a hostile nature and its colossal beasts.

One is thus forced to assume that, prior to all weapon and tool manufacture, the primitive human being had only, besides the tremendous power of his muscles and the agility of his limbs, his more or less animal-like teeth and nails as means of attack and defense.

The use and perfection of man-made weapons naturally resulted in a corresponding reduction in physical exertion and the softening of this natural weaponry. As the human being began to produce means calculated to provide protection and security as well as a relative degree of comfort in existence, thereby increasing his intellectual activity, the human physique, no longer compelled to unusual exertion and the show of force, gradually achieved a certain balance and poise. His predatorial traits receded to the same degree that his intellect emerged. The wounding and lethal features of his physical body were gradually displaced into weapons, giving way to a human appearance. The jaw receded into the organs of speech, the clawlike extensions of the hands he once walked upon became the protective nail coverings for his fingers as they worked. With demands subsiding, a convivial being with upright posture replaced the crudely modeled body formed for a savage way of life.

Of course we can only sketch a very broad-stroked outline of how these developments are supposed to have unfolded, based on inferences drawn from contemporary ethnography. Comparative geology has, on the other hand, established for us well-defined periods of time in which to group certain developments.

According to the general laws of development, for which there exist thousands of points of comparison the world over, Karl Siegwart, among others, assumes in his work on *The Age of the Human Race* that the human being is supposed to have passed through four distinct stages of civilization:

the crude, animal-like state of nature in which the savages lived is supposed to have lasted for millions of years; the half-savage way of life, marked by the advent of the Stone Age, he numbers at roughly one hundred thousand years; the period of incipient culture (passage from the higher Stone Age into the Bronze Age), which is characterized by the continuing development of forms of government, religion, and society, he estimates at seven millennia; and the period of higher civilization of one separate human clan (the ancient Greeks and Romans), is said to have existed for centuries.

Siegwart calculates the duration of the post–Tertiary Period and the known age of the human race at 224,000 years.[8]

Any pertinent attempt at working man out from his rude beginnings can proceed only on the basis of the theory of organic development. However, the prevailing conflict of opinion means that it is impossible at this point to decide which theory is better suited to doing so: either Darwin's theory as elaborated in *The Origin of Species* or a theory that instead assumes "original dispositions" unique to each species.

Among the works we will be pursuing here, Otto Caspari's *The Prehistory of Humankind* argues a brilliant position. The author is not partisan. After acknowledging the efforts of those who have earned their prominence as leaders in the field of the science of human beings on general ethnopsychological grounds, he introduces his own argument as follows: "I believe Darwin is owed great respect. It seems to me that I have only attempted to carry over into the study of the earliest spiritual life of humankind his renewal of a theory of descent in the study of natural history, through which he brought to light a fertile insight into the value of the history of organic development." Nevertheless, for Caspari, a thriving development depends entirely on whether "together we keep the ideal in sight and refuse to allow ourselves to be swayed by the spirit of a skepticism that recognizes in humankind no aims at all."[9]

From there, Caspari turns to his task: to lift the primitive human being out from the animal world and bring him toward the ideals of humankind, though not without again emphasizing the precondition of an "original disposition." Caspari explains the ascent of the human being from the animal in terms of the animal's absolute lack of the original disposition to manual dexterity and speech. "We have to recognize that the ultimate root from which particular factors of development are cultivated is to be sought only in dispositions of a spiritual and physical sort. All development originally proceeded from this inwardly rooted disposition to a splendid character and a temperament capable of being cultivated."

Now here we stand, with the human being before us, ascended from its original condition of unremitting defense against bloodthirsty predators, poised now to attack and destroy by means of apparatuses and *tools* manufactured by his own hand to increase the natural strength of the same.

Here is the actual threshold of our study: the human being, who, with the first equipment, the work of his own hands, discards his historical test piece to become the altogether historical human being, situated within the progress of self-consciousness. The human being is the only secure starting point for thoughtful reflection and for orientation in the world. This is because the human being is absolutely certain first and foremost of himself.

Occupying the center between the twin goals of research—the geological beginnings and the teleological future—is the human being: the fixed point from which thinking proceeds, forward and back, to expand the boundaries of knowledge, and to which it returns, in renewed health, from those regions to which research has no access because subjective interpretation has led it there astray.

CHAPTER 3

The First Tools

Now we must ask how the earliest tools and equipment were made and how, to a certain extent, they are still being made in tribal cultures today. In advance of an answer, let us agree on a few terms.

The Greek word *organon* designates, in the first place, a member of the body; in the second, its after-image, the tool; and in the third, the material, the tree or the wood of which it is made. In the German language, the words *organ* and *tool* are used more or less interchangeably, at least in the field of physiology, where, for instance, it makes no difference whether one refers to the respiratory organ or to the respiratory apparatus, whereas in the field of mechanics one speaks exclusively of tools. The more precise definition rightly concedes the organ to physiology and the tool to technology. In the internal organization of the body, the structures responsible for providing nourishment and conserving vital energy are called organs, as are those located at body's threshold that permit our sensible perception of things. Likewise, in the external organization of the body, the limbs, or bodily extremities should also be classed as organs.

Among the extremities, the hand counts as an organ in the strong sense, given its threefold determination: first, it is the human being's inborn tool; second, it serves as the prototypal image for all his mechanical tools; and third, because of its substantial involvement in the production of the material after-image, it is, in Aristotle's words, the "tool of tools."

The hand is therefore the natural tool, from whose activity the artifactual, the hand tool, proceeds. In all conceivable manner of position and movement, the hand supplies the organic archetypal forms after which the human being unconsciously patterned the tools he first required.

Composed of palm, thumb, and fingers, the relaxed, cupped, splayed, rotating, grasping, and clenched hand, either alone or along with the bent or extended forearm, is the common parent of the hand tools named after it. The first hand tools contribute directly in the manufacture of all other tools and all equipment generally and, as such, are the condition of their possibility.

The concept of tool ascended from primitive tools and expanded to include the tools of particular professions, the machines of industry, the weapons of war, the instruments and apparatuses of the arts and sciences. The concept comprises in the single word *artifact* the entire system of needs pertaining to the field of mechanical technology, where the human being alone has had its "hand in the game," whether in service to material necessity or to comfort and adorn.

As the human being makes use of the objects "at hand" in its immediate vicinity, the first tools appear as extending, strengthening, and intensifying the human being's bodily organs.

If, therefore, the natural hammer is the forearm with clenched fist, perhaps reinforced by a stone clasped in the hand, then the stone attached to a wooden shaft is its simplest artifactual after-image. For the shaft or the handle is an extension of the arm, the stone a replacement for the fist. Our reflections are limited here to a selection of figures belonging to the group of hammers, axes, and their nearest forms from the Stone Age, on the basis of their illustrative significance (Figures 1, 2, and 3).

This basic form of the hammer, in general capable of broad modifications depending on the material and intended use, has been preserved unmodified in, among others, blacksmithing and mining hammers and is recognizable still in the giant industrial steam hammer. Like all primitive tools, the hammer is an organ projection, or the mechanical reconstruction of an organic form with which, as Caspari tells us, the human being extends at will the natural power of his arm, intensified by the hand's dexterity.[1] Lazarus Geiger, in his work *Toward a Developmental History of Humankind*, expresses the hammer's significance as follows: "No matter how great the difference may seem between the modern steam engine and the most ancient stone hammer: one must imagine that creature who first armed itself with such a tool, who perhaps for the first time in this way

Fig. 1.   Fig. 2.

Fig. 3.

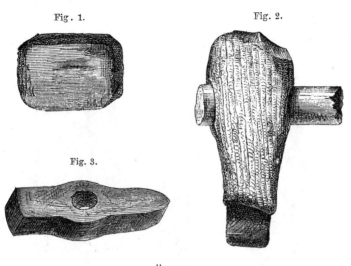

Hammers
Figure 1. Battering stone.   Figure 2. Stone hammer set in antler.   Figure 3. Stone hammer.

extracted a nut from its shell, was touched by the same breath of spirit that the inventor in our own day experiences as the flash of inspiration."[2]

Just as bludgeoning action is prefigured in the human fist, cutting action is prefigured in the fingernails and incisors. The hammer with a bladed edge is reconfigured into the hatchet and the ax; the crooked index finger with its sharp nail becomes, in its technical after-image, the drill; the single row of teeth turns up again in files and saws; while the clutching hand and the complete set of teeth are expressed in pliers and the vise-grip. Hammer, hatchet, knife, chisel, drill, saw, pliers are all primitive tools, "working tools" as it were, the first founders of civil society and its culture (Figures 4, 5, and 6).[3]

As the manufacture of tools continued to evolve according to the material used—whether wood, horn, bone, shell, stone, bronze, iron—the history of inventions resolved into a series of stages, corresponding with the favored material: the Wood Age, Stone Age, Bronze Age, Iron Age. Because the form is borrowed from the bodily organ, the stone hammer is just as much a hammer as is one made of steel. We are less concerned here with historical succession than we are with demonstrating, on the basis of available evidence, that the human being in fact displaced or projected the forms of his own organs into his primitive tools. What we

Fig. 4.   Fig. 5.   Fig. 6.

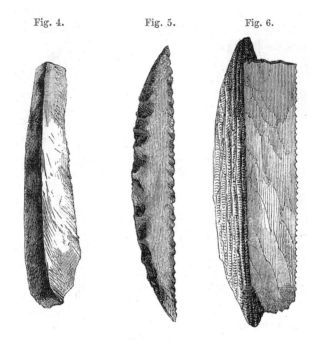

Blades and saws
Figure 4. Flint knife.   Figure 5. Stone saw.   Figure 6. Stone saw with antler grip.

want to emphasize is the inner affinity between the tool and the organ—
less a deliberate construction than an affinity emerging as a result of the
human being's unconscious groping—and the fact that, in his tools, the
human being is always only reproducing himself. Since the organ whose
practicability and power the tool is meant to potentiate is normative, the
organ alone can supply the form of the tool corresponding to it.

In this way, a wealth of creations springs from the human artifactive
drive and flows out through the hand, the arm, the teeth. The crooked
finger becomes a hook, the hollow palm a bowl. In the sword, spear, oar,
shovel, rake, plow, pitchfork, one can easily trace the dynamic tendencies
of the arm, the hand, and the fingers and their adaptation to activities
such as hunting, fishing, planting, and harvesting. As the stylus elongates
the finger, so the lance elongates the arm, augmenting its action of force
while at the same time, by decreasing the distance to the goal, also increasing the odds of reaching it—an advantage further compounded by
the momentum of the lance in flight.

The natural apparatus best suited to striking, tearing, and wounding is the arm that tapers into the fingertips which are equipped with nails originally very much like a predator's. Accordingly, pieces of wood and bone, whetted to a sharp edge or fine point, aided a natural motion and purpose. The skeletons of marine animals that washed ashore supplied the requisite material; inland, the bones of fauna as well as chert and flint could be found. At the same time, fire was also used in part to help harden, contract, hollow out, and polish these bits of wood and bone as well as to carve up stone.

The pointed end of a stag's antler or the jawbone of a cave bear could be used, exactly as found, to extend the hand, whose crooked fingers were not alone capable of loosening compacted soil (Figures 7 and 8). It may have been from such makeshift objects that the hoe was developed—which, representing the hand in the angle of the iron head and the arm in the wooden shaft, is *"a manner of appearance of the organ itself,"* to borrow a phrase of August Schleicher.

Fig. 7.   Fig. 8.

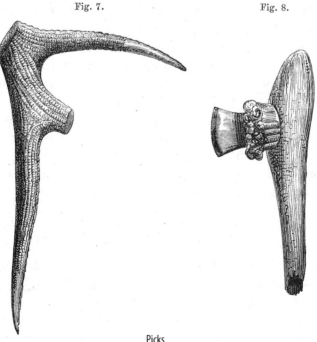

Picks

Figure 7. Pick made of antler.   Figure 8. Stone pick set in wood and antler.

These examples, drawn from a vast repertoire, sufficiently demonstrate that the elementary character of the tool is recognizable in all subsequent metamorphoses of the object.

Even the most complex products of advanced industry can disguise neither their source nor their essential significance. The steam mill and the primitive hand mill are equally apparatuses for grinding. The soul of each is and remains the millstone—the pair of boulders, one concave, one convex, comprising the grain-pulverizing apparatus that took over this function from the human molars. Throughout all its transformations, from water, to wind, to steam mill, the single component that makes the mill what it is—the millstone—remains; it does not matter if the stone is replaced by a metal disc.

In his chapter on "The Earliest History of Humankind in the Light of Language, with Special Reference to the Emergence of Tools," Geiger irrefutably shows that the tool's denomination is intimately related with an original organic activity, such that the word and the thing designated by it stem from a common root.[4]

The advantage of linguistic research, as we well know, lies in the fact that, like geological findings, those it yields can also substantiate as facts certain incipient and properly historical processes that are otherwise assumed *a priori*. For the roots of a language transsubstantiate into word families, both within a single language and from language to language. They are roots only in relation to the formation of trunks and branches, in which they live on. The root of a tree, cut off from the trunk, ceases to be a root; it is, in this extragenetic isolation, just another piece of wood. What is authentically recognized as a root in language is valid and lives on in it. And if, therefore, we can follow the traces of a tool's denomination all the way back to the activity of an organ that precisely accords with the use and purpose of the technical product, then, in this case, therein lies the proof that the technical product is produced and projected from the activity of an organ, and that, accordingly, its primitive form was provided and inscribed in the interior of the organism, in the unconsciously finding and outwardly re-creating artifactive drive. Just as the human being is always learning from the inside out, on the basis of his inner disposition, and he learns from the outside in only insofar as things provide the material for his power of representation, likewise the model forms for his mechanical formations flow from inside of him outward in his need to give form. Even the bodily organism, the embodied I-myself, is an interior vis-à-vis the not-I, the counter-I, into which the bodily organism expands and multiplies through mechanical reinforcement and extension of its limbs, albeit only to the extent that the limbs are normative for their mechanical substitutes.

According to Geiger, the human being possessed language long before he had tools and artifactive practices. "Consider," he writes, "any word denoting an activity carried out using a tool: we will invariably find that this activity is not the word's original meaning, that it previously denoted a similar activity that was carried out using only the natural, human organs. Compare, for instance, the ancient German word *mahlen*, to grind, or to mill in English, in Latin *mola*, in ancient Greek *myle* (μύλη). The ancient technique of grinding wheat or rye berry between stones is doubtless simple enough for us to assume that it has been practiced in one form or another since the primitive period. However, the word that we use today to refer to this tool-mediated activity proceeds from an even simpler intuition. The widely disseminated Indo-European root word *mal* or *mar* means 'to crush between the fingers' or 'to grind between the teeth.' This phenomenon, of naming a tool-mediated activity after an older, simpler animal activity, *is an entirely universal phenomenon,* and I don't know how to explain it otherwise than to insist that the denomination itself is older than the tool-mediated activity that it designates today, that the word already existed before human beings employed any organ other than those naturally provided in their bodies. Where does sculpture get its name? *Sculpo* is a variant of *scalpo* and at first only signifies to scrape with one's nails.

"We must beware of ascribing too great a share to deliberation in the emergence of tools. Invention of the first, most rudimentary tools almost certainly happened incidentally, like so many of the great modern inventions. Doubtless these primitive tools were not so much *contrived* as they were *chanced upon.* This view began to take shape for me particularly after having observed that tools are never named after the process by which they are made—never genetically—but always after the work they are meant to perform. Shears, a saw, a pick—these are things that shear, saw, and pick. This law of language appears all the more striking, given that equipment that are not tools tend to be named—genetically, passively—after the material of which they are made or the work that produced them. A hose, for instance, is everywhere understood to be the skin stripped, or *sloughed* from an animal. But this is not the case with tools and, so far as language is concerned, very likely never was from the start. The first knife can have been a sharp stone found at random and, I would say, playfully put to use."

In the same passage, Geiger also refers to shears to further illustrate the point. *Shears* denote a double blade, a two-armed cutting tool. The Indians and Greeks have a closely related word, and the Swedish word *skära* means "sickle." Before there were shears or shearing knives to aid in shearing their flocks, ancient Indo-Germanic shepherds would pluck the wool from their sheep. According to Pliny: "The sheep are not shorn everywhere. In some places, the custom of plucking still persists."[5] Given that *shearing* is closely related to *scraping* and to the Old High German name for that scraping animal, the mole, or *scëro*, it seems probable, in light of the basic meaning of the words to shave, to scrape, and to scratch, that shears were conceived as a tool for scraping and scratching the skin for the purpose of plucking hair from it. "In this way, we can think of the names of tools and the *activities carried out by means of them as having slowly emerged in the very gradual, ongoing evolution of a repertoire of human movements* of which the unaided human body was already capable from the start."

The lecture given on the ancient shearing knife by Professor Helwig at the end of 1874 at the German Archaeological Institute of Rome, where several models were on display, confirms Geiger's view of the sickle shape of the shearing knife generally and specifically corroborates the sickle shape of the primitive razor blade, which was known to have been used already in early antiquity.

Certainly modern shears proceeded from the sickle-shaped shearing blade, but what an interval there is between that ancient tool in the hands of nomads and the deftly wielded shears of Paul Konewka![6] Geiger attaches a great deal of importance to a distinction that is calculated to put a point on the whole truth in the expression *evolution*, as applied to the tool—namely, the distinction between primary and secondary tools. "*When observing its evolution, the tool wonderfully resembles a natural organ.* Just like the organ, it has its own transformations, its differentiations. One misunderstands the tool completely in wanting to think one has found the cause of its emergence in its immediate purpose. Klemm, for instance, has already pointed out that the drill is meant to have emerged from the primitive boring technique for starting a fire."

From the above, it is clear that in Geiger's opinion—whom even Hey-
mann Steinthal has heralded as the foremost researcher in his field, citing
Geiger's book on the origin and development of human language and
reason—the tool is, like language itself, the "absolute self-production"
of the human being. Language tells us which things exist and what they
are; but above all, as the human being's greatest spiritual self-affirmation,
language reveals what it is itself. In language, all things are sublated [*aufgehoben*], in the dual sense of the word—that is, not only preserved and
retained but also uplifted and spiritually transfigured. This, then, is why
we give such weight to Geiger's view with respect to the original formal
agreement of primitive tools and bodily organs that, with the help of
these tools, are able to move far beyond immediate contact with the
outside world.

An understanding of what Geiger has called the *evolution of the tool*
should enable us to give an account of the concurrent *evolution of the organ*,
which we would like to add to this. The hand of the primitive human being
was doubtless very different from the hand of the civilized human being,
to the extent that the use of tools made possible the latter's gradual
acquisition of a degree of tenderness and flexibility since the tool both
exercised and protected the hand. The hand was relieved of sustained,
immediate contact with raw, coarse matter, and the suppleness it required
to manufacture more complex equipment increased by means of the simple tool. This way, by reciprocal action, the tool supported the evolution
of the natural organ, the organ in turn supporting, at each correspond-
ingly higher stage of dexterity, the evolution and perfection of the tool.

The nearest stone or branch, found and seized as is by the foot of an
ape, remains a stone or a branch like any other stone or branch. But in
the hand of the primitive human being, the stone or branch becomes the
promise of the tool, the primordial cell of an entire cultural apparatus
yet to come. Merely selecting an object for a particular purpose brings
the human being nearer the concept of the tool. The chipping and whetting away of sharp edges and bumps from hard materials to enable the
painless handling of them should be considered the first free modification of natural objects.

This then properly opened the way for the fabrication of the first
tools, for the stone and the branch are the tool's embryos. Depending on
the form and quality of the chosen object, the branch becomes a staff,
a club, a lance, an oar, a bow, or a handle in a general sense. The stone

initially assists the hand in bludgeoning, cutting, drilling, whetting, polishing, and thereafter derives from it, in each of its subsequent metamorphoses, whether hafted in wood or set in horn, the first fixed measure and proportion. The tool becomes all the more *handy* to the degree that the essential characteristics of the creative hand, its form and motility, are embodied in it.

We alluded above to just how definitively the teeth and nails of the prehistoric human being—the weapon-like natural quality of which should not be underestimated—extend into the domain of primitive tool-making in the form of the chisel and wedge (Figures 9, 10, and 11). Cutting, sharpening, carving a point onto an object—these activities find their archetypal image in the teeth with which the primitive human being laboriously accomplished the tasks that he was later able to perform with much greater ease once in possession of the appropriate tools. In this study, we can go only so far as to hint at the reconfiguration of, for instance, the primitive hammer to a variety of purposes. For our concern is not with the history of tools; rather, our task is to indicate the significance of tool formation for the advancement of self-consciousness.

Fig. 9.   Fig. 10.   Fig. 11.

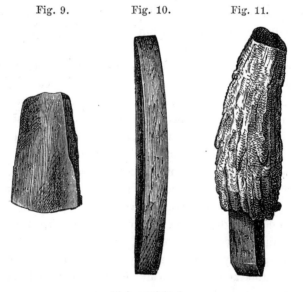

Wedges and chisels
Figure 9. Stone wedge.   Figure 10. Stone chisel.   Figure 11. Stone chisel with antler grip.

With respect to the Stone Age, the reconfigurations of the hammer must have been rather limited relative to the human being's meager ability to configure matter. With one side of the stone head left blunt to be used as a hammer, the other could be prepared with an edge to furnish the hand with a hatchet; the hatchet then could be enlarged to furnish both hands with an ax (Figures 12, 13, 14, and 15). Only later did metalworking enable greater diversity in reshaping the hammerhead. Whether

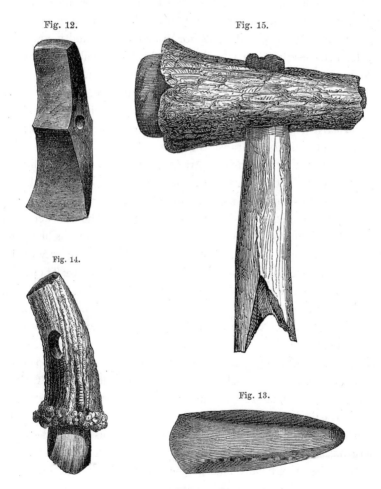

Fig. 12.   Fig. 15.

Fig. 14.

Fig. 13.

Hatchets and axes
Figure 12. Ax hammer.   Figure 13. Polished stone hatchet in antler.
Figure 14. Polished stone ax in antler.   Figure 15. Stone ax in antler, with handle.

pointed, narrow or broad planed, elongated, even pronged, and depending on the angle at which the head was attached, the hammer evolved into the pick, shovel, spade, pickax, hoe, and so on; whereas the hatchet, with a lengthened blade, evolved into the knife, the saw, and the file. Just as the sledgehammer was the simple likeness of the arm and fist, the different postures of the hand and arm served, as indicated, as the prototypal image for the corresponding transformations of the hammerhead. It will be left up to the reader to draw for himself, from the intuitions he may even have had as a child, further corollaries in the reconfigurations of these primary tools: how, for instance, stone and wood were transformed into projectiles, javelins and slingshots, ballistae and catapults, bows and arrows, blowguns, crossbows, rifles, and cannons; how the crooked finger of the hand that plucks became the sickle, the sickle the scythe, the scythe the reaping machine; and how the concept of an original activity, expressed in the tool's basic form, is preserved throughout the entire series of its transformations.

There is hardly a more visually striking example of such reconfiguration than that of a bowl formed in the likeness of the cupped hand. The bowl is the basic form of so much household equipment, its dimensions accordingly modified to produce the spoon, mug, pitcher, bucket, even the amphora and vase. The cupped hand's first substitute was likely the skin of a fruit, since generally those things nearest in similarity to the natural organs were the most convenient stand-ins for them. Thereafter followed the free working of materials: cups were carved from wood, molded from clay, hammered or cast in metal. In fact, the entire collection of the "Japanese Museum" in Leiden may be referred back to the palm of the hand!

To put it definitively—and the definitive voice seems often to come to us from research in the field of linguistics—we refer again to Geiger, from the essay cited above: "The use of tools that he has prepared himself is, more decidedly than anything else, a clear and distinguishing characteristic of the way of life of the human being. For this reason, the tool's emergence is a question of the greatest importance for early human history, and I therefore believe I am justified in framing the question as to the character of the early human being's equipment in this at once both limited and comprehensive sense. I do not hesitate to claim that there must have been a time when the human being possessed neither tools nor equipment but instead had to make do with only his natural organs; that thereafter ensued a period of time wherein he was already capable of recognizing and making use of found objects resembling these organs in order to increase, extend, and reinforce the power of his natural tools—to use, for instance, the empty skin of a fruit as a surrogate for his cupped palm, which had been his first vessel. It was only after he became accustomed to the use of found objects that creative activity entered his life, by way of reproduction."

It is important to remember that, at this point, the hand's immediate involvement gradually subsides. The hand tool rests completely in the human hand—that is why they are called hand saws, hand drills, hand files, hand hammers. With the machine, however, the human hand for the most part only specifies the start, direction, and halt of the movement. These mechanisms do not require the hand's direct and continuous contact. The hand alone guided the sickle and scythe; the reaping machine replaced the power of the hand with the power of the draft animal, supervised and guided by the human being. But no machine ever fully excludes the human hand; for even when a part of the mechanism detaches itself entirely—like the arrow, the bullet, or like the rocket-launched buoy rope for the shipwrecked—the departure is only seeming and momentary.

The stone grasped and held in the hand for protection and increased strength is in immediate union with the natural organ. The hand takes hold of it, and the arm performs the necessary levering movement. A combination of many movements takes place when throwing: the hand grasps and releases, the entire arm generates momentum, indeed the entire body takes part in flexion and extension. Not so with the machine! And although the machine, throughout its many stages of development, diverges ever further from the obvious external conformity with the bodily organ that it originally displays, this too lies within the concept of development. At the same time, though, this concept implies that an unchangeable element persists throughout all changes in configuration, from the stone that slew Goliath to the world-historical "shot heard 'round the world."

We will discuss further on in greater detail how, in designing machines, the human being had to refer unconsciously back to the prototypal image of his own living union of limbs, to transpose this into the lifeless machine parts in order to initiate in them coherent, purposive activity.

At ethnological museums as well as World Expos, the uncivilized world's simplest equipment comes face-to-face with the most complex machinery of modern life. The careful observer can identify the visible thread that runs throughout this story of development. According to two criteria—the nature of the materials primarily used for tools and the prevailing way of life—there are assumed to be distinct culturo-genetic periods, which tend to involve a clear ordering of need. One can distinguish a Stone, Bronze, and an Iron Age. It is also assumed that a hunting-gathering way of life preceded the period during which human life was organized nomadically around the herding of livestock. A period of time associated with permanent dwellings and the passage to art and science is thought to have succeeded this. One should absolutely refrain from positing a neat chronological sequence, however, even if only roughly demarcated. For there are nomadic societies that will never make the transition to a sedentary way of life, and there have been hunting-gathering societies that have taken up agriculture without ever having passed through a nomadic way of life. Likewise, the Stone Age remained completely foreign to the experience of the inhabitants of places where, in terms of raw materials, a surfeit of metal compensated for a lack of chert.

It is assumed that the preceding discussion, in presenting such tangible matters of fact, has at least begun to demonstrate that the first hand tool facilitated the organ's performance only through the closest possible formal similarity, and only in this way does it count as the homogeneous extension and reinforcement of the same. The complete demonstration will have to wait for our present discussion to unfold "one step at a time."

While the preceding discussion focused primarily on the form of the organ, and we only touched on its motion in passing, we will now turn our attention to the latter. In so doing, it should become apparent that the *organ's laws of motion*—keeping in mind that the primitive human being is as unconscious of these as he is of their transference into the after-image—impart a trace of the spiritual to the material that has been made to serve human purposes in tool-form. Hence, artworks and machine works both preserve the memory of their provenance—both in the organs of the human body and in the first equipment formed in the image of the organ. In this way, the human being maintains an inner relation with the artifacts belonging to the outside world that are produced in accord with the normative organs inside of him.

Striking evidence for this inner community into which the tool and its creator both enter through a principle of organ projection is found in the very first pages of Adolph Bastian's work on *The Legal Relations among the Various Peoples of the Earth*: "The human being who, like no other animal, is born helpless into the world, whom nature has provided no means of eking out an existence, is from the outset referred to artifice, to the fabricative activity of his spirit, in order to persist in his struggle with the environment. He fabricates weapons for hunting and fishing and *he will regard these products he creates as belonging to himself, and therefore call them his own.*" When Bastian then goes on to say that "the possession of weapons as such leads to his involuntary right to hunt and to fish as well as to the animals that make these activities possible," it appears then that the weapon—its possession and the fact of possession, which it is the first evidence of—is set in motion in concord with the human hand, as legal title, as it were, to the whole of living creation.[7] It is difficult to imagine a more inspired way of conceiving the intertwinement of the tool with the human Self. We cannot pursue this far-reaching idea in all of its implications, but only to the extent that its testimony is crucial for a theory of organ projection—since what is true of weapons is of course true of all tools.

Without exception, the hand immediately sets the whole hand tool in motion. The hand's involvement is what distinguishes the hand drill from the drilling machine. The hand tool's movement is the continuation of the movement of the hand and arm, as this movement is transferred into its technical extension, which takes the form of an accretion or attachment to the organ. This technical extension participates in the movements of the natural organ and complies all the more freely and easily the better the tool is adapted to the body.

The organic rules followed by the body's locomotor system are called, when applied to tools and machines, "mechanical laws." We should be cautious when using this expression, for the self-structuring organism is the work of its own absolute power, while the mechanism, brought into being by assembly from without, is a contrivance of the human hand. The organism, like the entire *mundo naturali*, is a becoming, while the mechanism is a finished thing. In the former, there is development and life; in the latter, composition and lifelessness. Those who disagree are blind to the difference between the corkscrew they carry in their pockets and the wrist as an integral participant in self-motivated organic activity.

Should the hand "concern itself" with an object for the purpose of performing a hoisting, cutting, knocking, or rotating movement, the object will cooperate with the hand—in whose grasp and command it is situated—in what it does, depending on the object's form and durability and the nature of the movement of the arm or hand. To say that the hand "concerns itself" with an object means far more than that "it grasps" or "takes hold" of it. The reflexive "itself" points to an agreement between the organ and the object appointed as tool. Even a baboon might grab a stone and toss it around, but its grabbing and tossing is fundamentally repetitive—the object is tossed away. To be sure, with each gesture of grabbing and even of tossing, the baboon gets closer to the human, to the tool; yet, every time it tosses the stone *away,* it relapses to an earlier state. But when the human being concerns "itself" with the stone and, through repeated grasping and inspection of it, begins to adapt it for use by his hand, then he is providing himself with it, arming himself with it. When he lifts the stone up off the ground he is already taking it into safekeeping for its proper use. And so, from this point on, the human being carries the stone with him and directs it as weapon and tool. The stone-throwing baboon repeats the same operation that it has for millennia, while the stone thrown by the primitive human was already the promise of the tool and an entire world of machines.

It follows, then, that if the hand that lifted the stone had concerned itself instead with a stick, then this stick would have been uplifted along with it into a lever; the sharp stone held in the hand cuts and pivots with it to become a knife, saw, or drill; for the sawing or drilling rotary motion of the wrist carries over into the object and shapes it into a knife, a drill, or a screw. Language designates the end of the lever as the lever arm, after its origin in the body. Just as grinding teeth came before the mill, lifting the arm precedes all levering activity. The performance of any tool has its origin in organic movement, and the original designation for an organic movement is the root of the name for the mechanism corresponding with it.

The characteristic relation of the natural organ with its mechanical reproductions expresses itself in terms of the so-called fundamental laws of mechanics. The content of mechanics is, as we know, the theory of equilibrium and the laws of motion. It would, of course, be inappropriate to directly apply the kinematic side of mechanics, as the theory of mechanisms of motion, to the movement of organic bodies, though it may prove invaluable as an aid in explaining organic motion. Matters of physiological fact always seem to leave something of a remainder that, in the context of purely mechanical laws, never quite add up. This is the prevailing difference between an organic and a mechanistic worldview in general and between the hand as tool and the hand tool in particular.

That being said, it should not come as a surprise that it is precisely in these primitive tools that organic movements have been so perfectly embodied that even the *nomen agentis*, the name designating an activity, has passed over into the object and become, in an eminently scientific sense, a physical term. The power of the arm to lift itself, decisive as this is for the concept of a variety of tools, is what gives the familiar *lever* its name, in much the same way that the hand's whorling and boring movements recur in the threads of the drill and the screw and the alternating momentum of the "dangling" arm prefigures the *pendulum*. Describing the "pendular motion of the arms," Georg Hermann von Meyer specifically emphasizes its linguistic derivation: "One arm *pendulates* backward while the leg on that same side is placed forward, and the other *pendulates* forward while the leg on that same side remains in place behind it; thus it is by means of this pendulating motion that an equilibrium is achieved, enabling the easy upright posture of the body, without too significant an oscillation either forward or backward."[8] Likewise Ludimar Hermann discusses the pendulum swing of the legs, plainly acknowledging the priority of the organic overall mechanism.[9] Applying the laws of mechanics to organisms for the sake of explanation surely does not make machines of them any more than machines are made organisms by transferring to them organic processes of motion. For, movement of the wrist signals muscle contractions, and these in turn blood circulation, nervous current, sense impression, conscious and unconscious representation, conscious and unconscious volition, ultimately sublimating in an I, whose absolute centralizing power manifests the harmony at play in the aggregate of organic activity.

The object of physiological study is the original—that which we have identified as the prototypal image for the hand tool. A good deal of time must have elapsed between the moment the primitive human being first appeared and the moment the first tool issued from his hand. And still even more time must have been spent in refining these tools before knowledge of the human body had advanced far enough to observe, as the discipline of physiology, this agreement between the quality, purpose, and successful application of the tool to the body's own constitution.

For the purposes of classifying things physiologically, the names and related indications of a number of tools migrated from mechanics back to their origin in the body. This is why in the mechanics of skeletal movement, expressions like *lever, hinge, screw, coil, axis, ligaments, threaded spindle,* and *nut,* for instance, have such a prominent place in the description of joints.

We should be sure to cite our physiological sources directly—in part to obviate any accusation that we are employing comparisons arbitrarily, and in part to provide the reader with compelling evidence of the naturalization of the language of mechanics in the territory of the organic. According to Wilhelm Wundt's *Textbook of Human Physiology,* the principal forms of skeletal movement are: "(a) *rotation around a fixed axis,* either as rotation around an approximately *horizontal* axis situated in the the joint, or as rotation around an approximately *vertical* axis almost parallel or coinciding with the axis of another moved bone. The first type are *hinge* joints, the latter pivot joints. A principal form of the *hinge* joint is the *screw-hinge joint.* The elbow is an example of this; (b) *rotation around two fixed axes.* To this category belong all joints in which the surfaces of the joint ends have a considerably different curvature in two directions perpendicular to one another. Either the surface in both of these two directions has a concordant curvature and differs only in degree (the curvature radius), or the surface is curved differently in both directions— that is, *convex* in the one direction and *concave* in the other. This second type of joint has been termed a saddle joint; (c) *rotation around an axis movable in a determined direction*—the *spiral* joint. The prototype for this in the human skeleton is the knee joint; (d) *rotation around a fixed point.* These joints allow for the greatest flexibility. To this category belong the ball-and-socket joints exclusively, that is, the hip and shoulder joints." Karl von Vierordt maintains that the *screw-hinge joint* is a subspecies of the hinge joint. "The clearest example of this is the screw form in the *tibia astragalus,* or the ankle joint. The astragalus, or ankle, represents a section of the threaded spindle, the surface of the tibia a nut with internal thread. The right joint corresponds to a left-whorled screw, and vice versa. Of course, the screw form is ordinarily only faintly indicated—that is, the size of the convolution, of which the joint forms only a portion, is very small. The elbow joint is another screw-hinge joint." It would also be worthwhile here to compare what Ludimar Hermann has written on *The Mechanics of the Skeleton* concerning "joints, *bonding* and *impeding* mechanisms, conditions of equilibrium, and active locomotion of the entire body."[10]

And so we can see how joint movements—especially those distributed between the shoulder and the fingertips—are integral to the tool's emergence and to the striking correspondence in names given to the prototypal image and to the after-image. Nor is this the only correspondence we find when we attempt to explain motion physiologically. There is also, for instance, the law of the parallelogram of forces, in connection with muscle displacement—a law that could not have been formulated if it had not been first organically realized. Form is given to the tool long before the formulation of the law that is only then later experienced and acknowledged as an unconscious endowment accompanying the form itself. For the nature of projection is that it is a progressive, primarily unconscious process of the self-externalization of the subject, whose every act is not always at the same time a becoming-conscious.

Just think of all the many processes and transformations that the original organic representative of the separating tool—the human incisor—has had to undergo, from the chip of stone resembling it to the sculptor's chisel and the threaded spindle, before the use of a sloping edge could expand our understanding of the decomposition of forces!

We should by now have effectively demonstrated that organic projection endows form to the tool. But that it also, at the same time, provides the formula for the action of force set to work will continue to occupy us, as we continue to demonstrate the truth of the maxim that man is the measure of all things.

The explication of this expression is the content of all knowledge. *Measure,* in this most comprehensive material and ideal sense, means no more nor less than *the typical ground of orientation in the world.*

CHAPTER 4

Limbs and Measure

Our investigation stands just at the threshold of the massive boom of modern culture. And from this standpoint, our concept of measure still does not quite extend beyond the domain of tools, in that it is limited, albeit provisionally, to an understanding of weight and measure in everyday life.

The foot, the finger and its knuckles, the thumb, the hand and the arm, the hand's width, the arm's span, the distance covered in a single stride—these are *measures of length*. The handful, the mouthful, the size of a head or a fist, the thickness of an arm, a leg, a finger, a thumb, a loin—these are *measures of volume and capacity*. The blink of an eye is a *measure of time*. A touch or a breath indicates something vanishingly small. These natural measures are still used, by old and young, primitive and civilized, to comprehend and describe the world. *"Unconsciously, the human being fixes his body as the standard of measure, even of nature,* and from youth onward he learns to apply this standard," remarks Gustav Karsten. "But now this method of assessing proportion that has become second nature for us is being done away with; and we are expected to undergo the learning process all over again. I confess that, though I work often with measure, whenever a metrical unit is specified I have the feeling of speaking a foreign language I have only inadequately acquired, meanwhile thinking in my native language and having to translate my thoughts. We will have to resign ourselves to this translation of our obsolete concepts of measure into metrical units; today's youth, however, will have to learn to think in metric."[1]

Throughout much of the world, the foot and the forearm were isolated as fixed measures of length, as standards or *norms*. When relayed to a body's surface and content, these also govern the measure of capacity, volume, and weight.

Since the fingers on one hand number five, from this is derived the number of digits on which the first counting systems were based. The ancient Greek word for this counting by fives was πεμπάζειν, or "fiving." Both hands together, totaling ten digits, supplied the decimal system, and ten fingers plus the two hands themselves the duodecimal system.

In a critique of Adolf Zeising's commentary on the golden ratio, Conrad Hermann remarks on the emergence of our counting system, that it did not arise of necessity immediately from number itself; rather, the numeral ten had to form the terminal or basic organizing unit of the entire system, and that this counting by tens at first appears merely as the human being's own subjective act of institution.[2]

Viewed in this way, one is tempted to imagine the *modus numerandi* as an invention, contrived on the subject's whim, that could very well have turned out otherwise—a misconception similar to that which long obscured our understanding of the emergence of language. The author is quick to add, however: "The immediate cause of this institution is given in the number of fingers on the hands, in that, experientially, these form the natural means or organ of counting for all humans everywhere, and the determination of numeral is initially bound up in this." This puts the question in the proper perspective. For the matter does not rest on a merely subjective institutive act of the human being, but on a universally valid and infallible organizational reserve, on the basis of which we may ascribe counting to ten on the fingers of the hand to absolute self-production. Hermann goes on, lending authority to our claim: "We believe we are permitted to assume that the ground of this organizational reserve is not merely inwardly subjective but also externally objective, and that the number ten itself can be attributed a certain decisive significance for the inner organization of the reality that surrounds us."

Where else could our mode of counting's rudimentary specification have been embedded but in the hand, from which tools and equipment and their dimensions proceed? Simultaneous with the tool, the hand projects the measure naturally inherent in it as well as its numerical values. The hand—the organ that comprehends material things and concerns

itself with them—is at the same time the organ that substantially facilitates spiritual comprehension and the discharge of representations. From
the inexhaustible store of its organization, the hand provides for the entire
world of culture. The old saying about the thumb making world history
is anything but a paradox; for the thumb constitutes the hand, and the
hand is the executor of the dictates of spirit.

Handicraft, commercial handlings, units of counting, weights and
measures, figures and calculations—all of these refer back to the hand.
Everything the hand accomplishes is, in the broadest sense of the word,
something "handled"—and this includes the manner in which the human
being handles himself. The hand is therefore involved in a profound grap-
pling, literally and symbolically, in the ethical domain: The same hand
that made the tool in its own image manipulates the tool as an instrument
of science, as a weapon of war, barters with the tool, passing it "from hand
to hand," molds it for artistic, religious, and scientific purposes.

Whatever the hand may be, it is not this in isolation but is this as *limb*,
as an organ belonging to a living, self-producing union of limbs—a whole
in which the least is preserved in the greatest and the greatest approaches
its truth in the least. The machine, assembled piecemeal of disparate components, clearly has *parts*; what it does not have are *limbs*.

We must wait until we are able to elaborate the hand's activities in
light of the concomitant array of powers available to the organism generally. At this point, so far as measure is concerned, what demands our
immediate attention is the foot as a standard of measure *par excellence*—
the foot as a norm.

With reference to the new international metric system, which means
to decommission the "foot," Karsten calls it a decisive mistake to abstract
the unit of measure so thoroughly from its familiar, age-old value, and he
acknowledges the legitimacy in "using the limbs of the body, for instance
the foot, as our, as it were, inborn standards of measure." I agree completely with Karsten. I would like to suggest, however, replacing the
restriction implicit in this "as it were" with an affirmation by saying that
the foot is our *actual* and *authentic* inborn standard of measure.

The foot has long served as a norm. And as long as there are human
beings walking around on two feet, the norm that is based on the length
of the foot will not cease to be. Nature and history are on its side—
the latter precisely to the extent that the metric scale, fetched from the

astronomical beyond, discloses itself, for all its exact scientificity, as fundamentally reducible to the foot's own enduring norm. An article published
in the *Magazine of Foreign Literatures*—"The Economy of the Ancients" by
French economist Jules du Mesnil-Marigny—demonstrates how firmly
the foot signals itself in spite of every disguise: "The natural system of
weights and measures is by no means a French but an ancient Greek
invention. The Greek amphora (= 26 liters), a measure of capacity, was
equal to the cube of one Greek foot (= 0.296 meters). The Greek talent,
the unit of both weight and coin (the latter in silver), was exactly as heavy
as 26 liters or one amphora of rainwater. The ancient system had the
advantage over the modern one, that weight and coin were based on the
same principle."[3]

It follows that one could have accepted the designation of a system
of weights and measures as the "natural" norm decreed by the bodily
organ, as opposed to the physical and astronomical standards based on
the geometrical partitioning of the globe.

The limbs of the human body are not only conducive, as we have seen,
to limiting the variety of specifications of measure, as these evolve at
different times according to local needs, but they also prove to be reliably
conventional units of measure the world over.

A tool—for instance, the hammer—is a solid and compact after-image
of the natural organ. A measure, by contrast, which is immediately available, represents only a single dimension of the body and its limbs. The
hand or the foot, when placed against an object, becomes the calibrat-
ing width of the hand or length of the foot. This technique, repeated
in a definite sequence along the length of a reed, stick, or rod, produced
the first measuring stick that, depending on the material's durability and
practicality, quickly assumed the handy form of a norm. If norms are
simply the embodied dimensions of an organ, then tools are a replacement for the organ itself. With the help of this replacement, the hand
manufactures more tools that begin slowly to depart from the proximate
formal identity that the technical after-image originally shared with its
organic prototypal image, often to the point of hardly evincing much of
a formal similarity at all. Yet, for all that, they are organ projections nevertheless. If anything, the projection is all the more powerful the more
essentially it reveals to us the organism's proportions and relations—
which themselves appear purer and more spiritually transparent the less

our attention is distracted by too great a fidelity of the plastic form to the natural prototype. The foot, as a norm that is quite remote from the form of the human foot, is a concrete abstraction of a single one of the latter's dimensions. This measure—having itself become a tool by virtue of its becoming a norm—assists in the manufacture of other tools, in the construction of buildings and machines.

So we see that one tool engenders another. The few forms of the primitive hand tool, on the one hand, and the incalculable diversity of civilization's scientifically sophisticated equipment, on the other, signal a concatenated progress—to which we can perfectly apply the theory of organic development, with its interlinked dependences. For instance, the gulf separating the measurement made by foot and the measurement made using instruments is clearly illustrated in an example taken from Wilhelm Foerster, with respect to the calculation of astronomical space and time:

"In a comedy by Aristophanes, Somebody is invited to dinner at the hour his shadow has a length of ten feet. If this length were referred to a pillar approximately the height of a man, then, irrespective of season, in Athens' climate the ten-foot shadow cast by the pillar would put us at roughly an hour and a half before sunset.

"Whereas today you would find Somebody consulting his watch to make it to dinner on time, back then you'd find him leisurely striding alongside this shadow."[4]

Foerster then proceeds to trace the most relevant and also apparently contrasting moments in the history of timekeeping devices—from the pillar erected in the ancient public square, to the sundial, hourglass, water clock, weight- and spring-driven clocks, the pendulum, and the chronometer—in order to show how the science and art of measurement proceeded to develop, from pacing the length of a shadow to appending and recombining mechanisms that we already know to be projections of organic powers (levers, springs, pendula, etc.). He shows, moreover, how human beings had begun to measure calendrically not only astonomical time and space but even sensations and the formation of representations—a metamorphosis of the primordial human measure that really does border on the miraculous!

With measure and number, the human being recognizes and presides over things. Pliers, a primitive tool, aid in gripping something and holding it secure; the animal's claw does this of necessity. But, with standards of measure and number in hand and eyes fixed on the clock to secure time and space in the *calendarium*, the human being arrives at his highest undertaking—to be one who *measures*, who *assesses* and *thinks*.

This is how the foot—the symbol of autonomy, the organ enabling self-support and self-propulsion—became a tool, became a norm. This is how the fingers arranged on the hand determined the mode of counting—which, as such, was not expressed in a tool *per se* but as a denary metrical scale impressed in tools of assorted purpose and inscribed in ciphers.

CHAPTER 5

Apparatuses and Instruments

Our reflections to this point have ranged within the sphere of the body's extremities, the hand and the foot. We now turn to the organism's semi-extremities, to the sense organs that, intermediary between the external world and the interior neural world, are situated at the threshold of both. It is *vision* that is in closest contact with the extremities we have already discussed, because measure and number are submitted to its immediate oversight. The eye is the organ of light and the prototypal image for all *optical apparatuses*.

Friedrich von Hellwald writes of the earliest and simplest form of an optical tool: "Little by little, the Babylonians, the Egyptians, and the Greeks all got to the point where they were able to calculate the distances between heavenly bodies and earthly continents on the basis of geometrical forms that they believed to be inscribed in the heavens themselves. This is the point to which sense perception had been brought centuries ago, just by observing the stars through a simple elongated tube— not unlike the way we might inspect a painting in the museum today through a rolled up sheet of paper."[1]

These tubes were nothing other than the mechanical extension of a hand forming a tubular shape that a person involuntarily held up to his eye, for want of another aperture. The hand aperture itself is an enhancement of the natural shadowing apparatus existing within the eye. The extending tube formed of curled fingers amounted to the most primitive *telescope*, the improvement of which followed upon the invention of glass and the art of grinding it.

By that point, haphazard incidents had already led to the perception that small bodies appear larger when viewed through a spherically shaped piece of transparent glass. In his history of physics, Julius Zöllner remarks on Seneca's familiarity with the magnifying properties of glass globes filled with water. The intricacy of ancient Greek stonecutting would seem to indicate the use of magnifying lenses, though the "lenses" that have been unearthed may well have been used exclusively for lighting fires, since the sacred fires of Vesta could be lit by no other source than the sun. In the mid-eleventh century, the Arab Alhacan (Ibn al-Haytham) was likely the first to put actual lenses, made from a segmented sphere, to use as magnifying glasses. "Their use was limited: they were laid directly on top of the object itself, for instance on top of the letters of a document."[2]

It was the art of grinding eyeglasses, begun in the thirteenth century, that led to improvements in lens making. Every converging lens is at the same time a simple magnifying glass, or *loupe*. The earliest such devices—commonly called flea-glasses *(vitra pulicaria)* since they were used to examine small insects—have since been improved to become instruments in the proper sense. Today's compound *microscopes* "are destined," in the words of Gustav Jäger, "to complete our knowledge of the organic world."[3]

Through all its transformations, from the simple magnifying glass to the solar and oxyhydrogen gas microscopes, the lens remains the single constant, the soul of the instrument. The spherical piece of glass unconsciously formed on the model of the crystalline body of the ocular bulb acquires its name from the general formal similarity it shares with the bulbs we know from gardening (Figure 16). But the riddle of this organ's physiology could be solved only after having projected itself as an organ of sight into a number of mechanical apparatuses, in this way establishing the external reference for its own anatomical construction. The human being has taken what he has learned from the instrument he unconsciously patterned after the organic optical tool and conveyed this to his understanding of the actual focusing of refracted light in the eye, to the "crystalline lens."

The same holds for the *telescope*, invented in the early seventeenth century by the Dutch lens grinder Hans Lippershey and later developed into the giant telescope by William Herschel. Its most basic component is also the lens. Despite the instrument's variety in form and composition, the concept of light refraction remains constant.

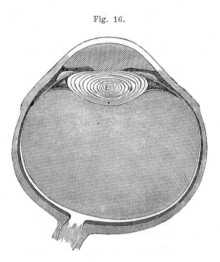

Fig. 16.

Figure 16. Eye with collecting lens.

The term "lens" is an example of the lesson we can take from anatomical and physiological nomenclature generally: the names we give
to organic bodies are largely derived from objects external to them, and
especially from objects associated with the projection. How else should
we understand it, when the construction of the eye is found to be *entirely
analogous* with that of a *camera obscura*, when it is shown that an image
of an object in front of the eye is formed inverted on the retina *"in exactly
the same way"* as the image forms on the back panel of a *camera obscura,"*
and when the eye is alleged to be an organ that *"carries out with extraordinary precision the *daguerreotype process"*?[4]

From the standpoint of organ projection, one need only reverse these
claims and explain that the *camera obscura*'s construction is entirely analogous with that of the eye, that it is in effect the unconsciously projected
mechanical after-image of the same, by means of which science is belatedly aided in comprehending visual processes. This is exactly the claim
that Carus is making when he writes: "In actual fact, one could hardly
suspect that the eye required an image to form on the retina before the
daguerreotype was known; *for this discovery first allowed us to conceive* how
rapidly and with what extraordinary variation and freedom the action of
light can permeate a substance."

Carus—acknowledging in general that "the artificial apparatuses used
to enhance eyesight (telescopes, microscopes, eyeglasses) demonstrate use

of the laws of optics, as they are given perfectly in the eye"—concludes his thoughts on the organ of sight with an analysis of the *achromatic property* in the eye's refractive medium. He offers incontestible evidence that the eye is unconsciously reconstructed in the optical apparatus, because to think an optical apparatus is realized in the eye would be an incorrect and derivative point of view. The circumstance he is suggesting is inestimably important for us, because the concept of organ projection—as a long unacknowledged scientific principle—is unconditionally confirmed in it, beyond the slightest doubt.

The following are passages excerpted from Carus's *Physis*. They deal with the elimination of visual disturbances produced when observing an object through a simple glass lens, specifically the colored halos that tend to form around the observed object. Since a lens is present in the eye, which functions as the collecting mechanism on which our vision depends, it was more or less expected that in the eye these same colored halos would also be produced. But optics teaches us that when colors are produced simply by looking through a prism, these colors can be neutralized when a second prism is placed over the first and the gaze has to pass through both together. This is how the idea came about that colored halos could be prevented from forming in the telescope by assembling the lens of *two* separate glasses—a crown and a flint glass—which solved the problem perfectly (Figure 17).

Fig. 17.

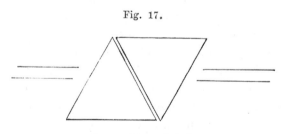

Figure 17. Achromatic device.

"*What artifice accomplished only through considerable detour, the unconscious formative power in us perfected long ago* in joining with the lens the larger, slightly more viscid vitreous body of the eye—the lens acting as the quasi-second prism neutralizing the colors produced by the first to achieve a clearer image.

"I admit I have always thought of this ingenious apparatus in the eye that can effect an achromatism (elimination of false color) as one of the most remarkable instances, making explicit how all of our conscious scientific practices must actually have in mind the same law that human physis has in reality been persistently and perfectly following in an unconscious manner, long before we could possibly know anything about it."

Surprising facts of projection similar to those in connection with the eye, as the organ of light, emerge in connection with the ear, as the organ of sound. The eye and the ear are the preferred senses of our intelligence. Their deeply interior affinity is expressed in such phrases as a *color's tone* and a *tone's color,* thus explaining our transition directly from a discussion of optical to one of acoustical technologies.

As with the eye, the cupped hand was also the first tool applied to the ear to enhance its natural functions. The ancient astronomer's "looking-tube" corresponds to the *ear trumpet,* the conical funnel extending the ear canal. This tool acts in part, like eyeglasses for the visually impaired, to help the hard of hearing, and, in part, like the stethoscope is used for medical diagnoses, to enhance a healthy auditory function. Just as the arm and hand extend into the hand tool, the extension of the ear canal and outer ear was originally designed for a conscious purpose—albeit without awareness of the analogon existing between the design of the instrument and the physiological processes taking place within the organ.

As the science of acoustics advanced, epoch-defining inventions appeared, as they did in the field of optics: in the latter, these were the *daguerreotype* and the lens with its many combinations; in the former, these were the *monochord* in its many forms and the *keyboard instruments.*

These unconsciously occurring discoveries were made long before science recognized them as projections relating to the activities of the senses. They are therefore only *post factum* explicable in terms of organic processes. The significance of mechanical apparatuses increases with insight into the diversity and scope of physiological processes. Investigations into the processes particular to each sense organ, according to Hermann von Helmholtz, must address three different aspects: "a *physical* one: how the agent that excites sensation—which, in the ear, would be sound—reaches the sentient nerves; a *physiological* one: investigation of the nervous excitement that corresponds with different sensations; a *psychological* one: the law according to which representations of specific external objects, in other words perceptions, are produced from these sensations."[5]

After completing his work on *Physiological Optics* and on the *Sensation of Tone*, each an intensive ten- and eight-year task, Helmholtz more or less put a provisional end to investigations into these processes.

Under his scientific eye, the "musical instrument" became an "instrument" in the superlative sense—that is to say, a tool through which insight into the organic substrate of the operations of spirit could be released: music had been transposed into acoustics, science explained and transfigured through art.

On the *monochord*, antiquity had discovered the consonants for tones. The simple apparatus consisted of a single string pulled taut across a resonant wooden box. A bridge was placed beneath the string, and this could be moved in specified intervals along the length of the string. The different ratios of string length seemed to determine the pitch (Figure 18).

Fig. 18.

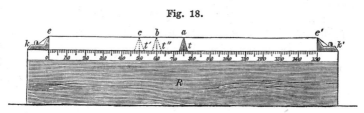

Figure 18. Monochord.
Resonance box *R* with rule for adjusting the sliding bridge *(t)*. The instrument's only string *ee'* horizontally spans the upward-bent, sharp-edged supports *kk'*.

So long as one was ignorant of the oscillation frequency and periodicity of sound waves, the tonal ratios Pythagoras established were regarded as a profound mystery. "In that modern physics shifted its focus from the length of strings to vibration frequency," Helmholtz solved the ancient Pythagorean riddle and referred the tones of all musical instruments to this solution.

The monochord represents the first in a series of stringed instruments of ever increasing complexity that culminates finally in the keyboard. At the South Kensington Museum, in a previous exhibit similar to this year's, a complete collection of musical instruments was on display, including a category consisting exclusively of stringed instruments, which afforded a coherent overview of this ascending series all the way up to its currently perfected form. It was the modern grand piano that provided Helmholtz the key to unlock that two-thousand-year-old riddle ensconced inside the inner ear.

Even earlier than Helmholtz, Alfonso Corti discovered in the cochlea a microscopic structure composed of some thousand or more fibers of unequal tension and length. This structure is now known as the organ of Corti. According to Helmholtz's investigations, the organ forms "a sort of regularly graduated set of strings such as one would find in a harp or piano" (Figure 19).

Fig. 19.

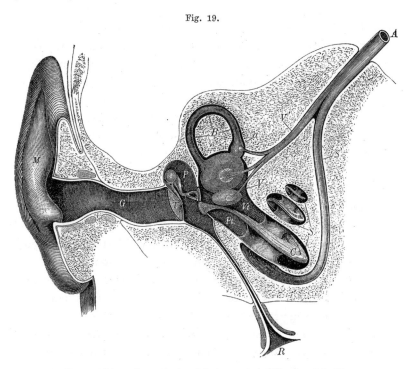

Figure 19. Schematic cross section of the human organ of hearing, right side.
M outer ear; *G* outer auditory canal; *T* ear drum; *P* tympanic cavity; *R* the Eustachian tube;
V, B and *S* the bony labyrinth, *V* the vestible, *B* a semi-circular canal with its ampulla *a*, *S* the
cochlea, divided by the spiral limbus into the vestibular canal *(Vt)* and the tympanic canal *(Pt)*;
A the stem of the auditory nerve, entering the auditory canal and splitting into two main branches
(V' and *S');* *S'* the cochlear nerve, entering the canaliculi of the central columella of the cochlea in
order to arrive via the bony spiral limbus at the organ of Corti *c,* which sits on the upper or
vestibular canal surface of the membranous spiral limbus.

"The organ's approximately three thousand fibers, each of which is tuned to a different pitch, correspond to the strings of a piano. Each fiber is affiliated with an acoustic nerve. The nerve is mechanically stimulated and perceives one specific simple tone as soon as the affiliated fiber is moved in sympathetic vibration.

"However: while piano strings vibrate sympathetically *only when affected by tones with which they share a harmonic likeness*, the fibers of Corti vibrate sympathetically *only when,* after having passed through the labyrinthe fluid, sound waves reach them whose vibration frequency pertains to *that particular tone* to which the individual fiber is precisely tuned."

I highly recommend Johann Czermak's essay on the ear and hearing in his *Popular Lectures in Psychology*—from which the above passages as well as Figures 18 and 19 are taken. I recommend it for the clarity of its presentation to those who are less familiar with the more comprehensive *Sensations of Tone,* notwithstanding that Helmholtz's description of the organ of Corti has recently undergone some revision.[6] At any rate, a truth once veiled in the obscurity of an unconscious idea has emerged into the daylight of modern microscopic science. The monochord is the incipient piano; it is the key to the harmonics of antiquity, made conscious in the theory of the *sensation of tone.* In his work on the *Fundamentals of the Esoteric Harmonics of Antiquity,* Richard Hasenclever offers a new take on an old question, arguing "that the natural basis of the harmonic system, by force of inner necessity, had to have been one and the same, for everyone, everywhere, for all time."[7] Instead of engaging a problematic, mindlessly self-regarding mysticism, Hasenclever helps us to understand that a profound mystery always conceals a great truth— a truth whose nature it is, as such, to approach the restless human drive for knowledge, even in veiled form.

By virtue of the eternal self-sameness of natural laws, "a possible explanation for Pythagorean harmonics has been found, due in no small part to a better application of the exact natural sciences, the physics of acoustics in particular." But, dilating out to a general aesthetic theory, the explanation also appears as a cosmically enmeshed propaedeutic to our knowledge of the physical and the moral world order. Because wherever human nature operates, the human profit on his effort and labor of thought is always the human being himself. And with this we can deduce the deep import of Hasenclever's question: "Did Pythagoras stop there, with that elegant finding presented to him in the monochord and with what this immediately implied for the construction of a tonal system? Or was the enigmatic province of sound disclosed ever more deeply to his investigative spirit, until at last he reached the limit where the causes of real appearances could only be grasped and understood on the basis of a higher principle, transcending all purely physical conditions?"

It should be evident how important the matter of the monochord is. But there are still further implications to be drawn from this tool of sound as a tool *per se*. It's likely this agreement between the self-projecting organ and the projected tool had been suspected earlier, though the suspicion did not develop beyond a vague and abstract sense. It was in the same metaphorical sense in which one speaks of a body of water as the eye of a landscape that one spoke of the elbow or shoulder joint as a lever—without considering that here we are dealing instead not with likeness but with an inalienable identity, to the extent that one side of the comparison is the after-image of the other and that, moreover, this fact completely rules out any other way in which the after-image could have come to be. Our everyday notion of the figurative perishes in the real image and after-image. The oar is the strict after-image of the outstretched arm and palm of the hand. We need to put an end to the tendency to diminish this reality through simile and metaphor, with a "quasi" or an "as it were." This is not an issue of the arbitrary selection of one from any number of possible explanatory figurative similarities, but what we are seeing is the singular and necessary similarity—or rather, consubstantiality. Where else in the world—besides in the inner relation prevailing between the mechanical lever and the organic levering apparatus consisting of muscle and bone, between the lens and the crystalline body in the eye, between the daguerreotype and the ocular bulb, between the pendulum and the swaying arm, between the screw and the rotary motion of the hand, between the stringed instrument and the organ of Corti—where else in the world does one find a second such agreement between a mechanical apparatus and some other prototypal image existing *outside* of the bodily organism? This agreement will of course not be so clearly evident in everything we discuss, and much will be made apparent only in the course of time. After all, centuries transpired between Pythagoras and Helmholtz, between the monochord and the "piano in the ear." The human body is universally coordinated with the cosmic

conditions of its existence or, in other words, with the forces of nature—
the eye with light waves, the ear with sound waves, the lungs with the
atmosphere, the intestines with nourishment, the nerves with electricity,
and so on. Within these coordinates, the aggregate movement of the
organism's limbs and the aggregate functioning of its organs recur, sensuously expressed in matter, in one of the countless works of technology
that correspond in endless variety to the body's organization.

It is not only in this world of substance, but also in that other world
toward which our investigation urges us, in the world of spirit where we
are made, as Carus writes, "fully conscious of a spiritual organism exhib-
iting itself in a unity of substance with a palpable build." In both worlds
an analogy proliferates, its logical *moments* admittedly less sensuously
comprehensible than are the *movements* of hard matter, though manifesting teleologically with greater permanence and transparency the more
the energy of our self-consciousness increases.

Accordingly, this play with fantastical and arbitrary comparisons on
the order of "quasi" and "just like" and "as it were" must give way to the
fact of organ projection. For it is on this basis that the theory of tonal
sensation discovered the fibers of Corti, and it will be on this basis that
further research into the "labyrinth" of hearing will be kept from going
astray.

Wilhelm Wundt remarks in the chapter in his textbook on the hearing
apparatus that, according to Victor Hensen, it is not the fibers or the
bow in the organ of Corti but the basilar membrane that is attuned to
specific pitches, depending on the variable width of each section. Helmholtz corroborates this hypothesis in the third edition of his own work,
though he believes that the organ's bow, as a relatively fixed structure, is
specified to convey the vibrations of the basilar membrane to the narrowly
circumscribed region of the nerve bulb. His innovative idea, of approaching the problem by way of the piano, remains unchallenged.

The theory of tonal sensation is in fact so widely accepted that it will
be to our advantage to cite further from the pool of knowledgeable expo-
nents of it, as a kind of guaranty or preemptive defense against whatever
doubt or opposition there may be to the theory of organ projection:

"In fact, the vowel sounds that resonate from the piano are similar in
aspect to a resonance concerto. And couldn't one also imagine the fibers
in the organ of Corti as a *harp in the ear*, which is moved by resonance to

sympathetic vibration? Resonance provides the means for the analysis and synthesis of sound."[8]

"One branch of the auditory nerve is in contact with an astoundingly delicate apparatus, the cochlea, *a complete keyboard* composed of several thousand keys that is used to recognize tones and the sounds composed of them."[9]

"The orderly arrangement of teeth in the organ of Corti, which waves of water pass over like the keys of a piano, tempt one to imagine that each key, each tooth is in a sense tuned to a particular pitch."[10]

"The auditory nerve picks up these movements by means of quite peculiar, *formally attuned fibers*, such that one specific fiber will only ever be excited by the tone specified to it, the particular sensation of tone being based on that tone's appearance."[11]

"It seems, then, that the latest puzzle presented to researchers by the structure of the ear has at least hypothetically been solved. In support of Helmholtz's view, we may as well mention Hensen's discovery that in some species of crab, tiny exposed hairs coupled with auditory nerves are set in motion by the sound of a bugle in such a way that each distinct tone causes a distinct hair to vibrate."[12]

"As the fibers of the organ of Corti gradually decrease in length, corresponding to the narrowing of the spiral limbus from bottom to top, they form a kind of graduated set of strings, like we might find on the *harp* or the *piano*."[13]

Thus, we are convinced once again that a mechanism assembled of component parts by the human hand may be built in the most striking conformity with an existing organic structure, without the least knowledge of how the latter works. Given this absolutely singular identity, the fact is that whatever served for the human being as an unconscious prototypal image later assumes priority, by means of its reproduction in the after-image. Recall that the ear drum, the tympanic cavity, the Eustachian trumpet already took their names from musical instruments (even the cochlea recalls the violin's scroll); so it no longer comes as a surprise to us to hear Helmholtz's great discovery spoken of quite naturally as a harp or piano in the ear. In this way, we introduce a provisional representation of something that even the layperson can understand. The observation bears repeating, to what extent language's need of image winds up conveying, with divinatory clairvoyance, so many names from

artifactual tools to qualities and processes of organic activity. Anatomical and physiological terminology is just as much the counter-image of
technical terminology as the material products of technology are the
image of the organ.

Looking back at the tools we have discussed so far, it seems we have
been primarily concerned with the first and the final manifestations in
the course of their development. Between the two, one finds a myriad
of variations in form that is difficult to assess. To take these even roughly
into account would take us far beyond the limits of our investigation.
Moreover, we are still so sorely lacking in adequately established physiological materials that, for the great majority of these transitional forms,
the organic prototypal image could hardly be substantiated in any satisfactory way, certainly not as thoroughly as we were able to do with the
preceding examples. It would also be superfluous for us to go into greater
detail in those instances where the facts are already so well established,
attesting to the profound affinity between the nature of the union of limbs
and the essence of the tool's emergence. The concept of organic self-
activity implies that this affinity is the enduring expression of difference;
and that, as the ear conveys only sound waves and the eye only light waves
to the locus of representation, the various representations are transposed
into hand movements, and these hand movements then form tools and
equipment—in other words, *they project* in concert with themselves.

And so it turns out that the astonishing facts that are brought to light
through scientific research and that we may invoke with genuine confidence in their applicability to organ projection have an evidentiary force
extending well beyond the individual case. For, in the field of the organic,
the individual case is always significant: like the fossilized bone from which
the entire skeleton of a previously uknown prehistoric creature can be
reconstructed, the individual case guarantees for all cases appertaining
to it the same certainty that it asserts for itself. In fact, the individual case
constitutes the whole so fundamentally that the whole itself stands and
falls with the individual!

A good deal of world history had to elapse before human effort and
research succeeded in discovering the piano in the ear. Only a few decades
ago, Carus believed we would never be able to explain how, in the small,
extremely tender soft tissues of the ear's labyrinth—hardly the size of half
a hazelnut—all the millionfold sound vibrations instantaneously alight

to produce, via the nerves, perceptible sounds. And today it is thought impossible that we will ever be able to explain how, in the daguerreotypic layer of the retina—no bigger than a pea—we have absorbed the firmament in all its splendor.

Still: see how our knowledge has advanced and think what a world of discoveries still awaits the investigative drive that revealed to us the hallowed mysteries inside the ear!

One dare say *a world of discoveries,* in view of that subtle flash of contact between sound vibrations and delicate nerve endings, through which cosmic movement is translated into spiritual movement, the magnitude of stimuli into the magnitude of sensation and representation, through which matter is translated into thought!

The progress we have since made in our knowledge of organic life—no matter how meager this seems in comparison with the tremendous task urging us on toward a distant horizon—is of even greater weight, since, as the stable pillar on which our previous discussions are based, it holds in itself the prospect for that which is still to come. After all, it comes down to the insight that *all* organic structures, from hard bone to soft and delicate tissue, are destined in one way or another to project themselves outward into human contrivances—in both the most common and the most sublime sense of the word—in order to be employed retrospectively as scientific apparatuses to increase self-knowledge and knowledge in general.

This retrospective method of scientific research proves remarkable with respect to the *vocal organs,* to which we now turn our attention, given their immediate rapport with the organs of hearing—which physiological evidence indicates were originally formed out of the respiratory organs.

As the daguerreotype and the piano elucidated the operations of the eye and the ear, research into the vocal organs was aided by the pipe organ, as their mechanical after-image—an after-image so faithful and precise that comparing it with the original elicits a smile of astonishment in the observer not unlike a strikingly lifelike portrait also might.

The vocal organs, as we know, consist of the chest cavity, containing the *lungs,* the *trachea* and the *larynx,* and the *pharynx* that opens into the *oral and nasal cavities.* We recognize their projection in the corresponding principal parts of the *pipe organ:* the *bellows,* the *windchest,* and the *pipe* with its *resonator*. Physiology needs these parallel structures in the musical organ to explain the functions of the vocal organs, and it also has recourse to terms borrowed from other musical instruments as well, when describing organic structures.

"Oddly enough," writes Carus, "it is the most solemn and massive of musical instruments—the organ—that offers the richest comparison here, but with the difference that the *single* living, moveable pipe in the organism—the trachea, equipped with a glottis that voluntarily expands and contracts (the English literally call this a *windpipe*)—takes the place of the many immoveable pipes of various size in the organ."[14] On this note, Czermak's illuminating treatment of the anatomy and physiology of the vocal apparatus leaves little to be desired. He explains "the exact nature of the movable chest cavity in which the lungs are entirely enclosed, as a sort of *bellows*"; he speaks of the "windchest, in other words the trachea and the two bronchial tubes"; and he provides an account of the larynx and its outlets, "notably, the single pipe with its resonator, in other words, its vocal tract."[15]

And yet, some will continue to insist on the contingencies of resemblance, where indeed we are dealing with nothing other than an organic order and an inner necessity that language also confirms.

Czermak's lecture on *The Heart and the Influence of the Nervous System on the Same* is no less conducive to an understanding of organ projection.

Beginning with anatomical observations about the external form and the internal structure, he proceeds to explain the mechanics of the heart in terms of a *pumping station,* in other words, to explain the manner in which the heart, by alternately contracting and relaxing its four chambers and through the play of its *valves,* drives the blood to circulate throughout the body's vessel system in a particular direction. His deliberate and repeated use of the term *heart-pump* leaves no room for doubt that by choosing precisely this word he is corroborating the clear conformity between the inborn and the artificial tool.[16]

This compound word is, in a sense, the most concise abbreviation of our basic approach: the organ and the mechanism are combined in a single concept, together with the myriad interrelations that emerge in the process of projection! But, while we welcome this tacit endorsement on the part of language, we should not fail to address the weaker side of the same.

When we say "heart-pump," we are not taking the sort of metaphorical license implied in expressions like "eye-catching," "earmark," "wholehearted," and so on; instead, we are taking the part of the compound designating the machine ("the mechanism of the heart as a pumping station," in the example above) absolutely literally. Otherwise, the expression would be yet another instance of an already out of control terminological imprecision in the exact sciences, one that conflates the organic and the mechanical and has resulted in a serious confusion of concepts. An organ is never a machine part, nor is a hand tool the limb of an organism. A mechanical organism would be tantamount to something like organic clockwork—a squaring of the circle. In what follows, we will find ourselves having to insist time and again on the necessity of a conscientious sorting of what one designates as mechanism, though at this point we can conclude with the following provisional remark. We can account for the confusion of concepts in the essential relations subtending the process of projection as a sort of involuntary and unremarked slippage of the representation of the organic from the prototypal image to its mechanical after-image, and vice versa: when the machinal is used to explain organic processes, then, in the fervor of experimentation, the mechanical slips into the organism unremarked, such that impermissible confusions necessarily arise alongside problematic figurative explanations.

Preventing or correcting this tendency used to be only a matter of attentiveness, as long as the tools were simple or, when assembled, easy to surveil. One did not presume, for instance, to speak of "the organism of the carriage" or of "the mechanism of thought." Meanwhile, however, since complicated machinery and apparatuses have superseded work by hand, a remarkable confusion of the concepts "mechanical" and "organic" has become habitual. Moreover, the manufacture of machinery required for large industry and worldwide communication necessitates such expenditure of scientific technology that, in our ignorance and absentmindedness, we fail to notice the transfer of motive powers, and the steam engine and the telegraph assume the semblance of self-generated activity.

Under the spell of this semblance, the beholder feels he is confronting something kindred, not an imaginary but an actual demonic power emerging from the mechanism. And what else could this power be but a reflection of something spirit-like, such as manifests in relations that the human being abstracts from organic life and represents in the form of the laws of mechanics? Vis-à-vis the eternal flux of relations in the world-organism, these laws are an expedient to representation; they represent relations among the living, torn from the cosmic context en route to abstraction; they become, thus out of context, frozen in signs and numbers and combined into formulae to be used as tools of thought—the concept of tool meant here in the broadest possible sense.

Of course, physical laws perfectly suffice the mechanism, but not the organism, which we are able to grasp only insofar as we are coextensive with it. Whatever exists beyond this is the great mystery of life that the mechanistic worldview absurdly claims to have solved. At the very most, we can accept the intentional permutation of the concepts mechanism and organism in the manner described above as a semantic strategy, on the condition that we remain aware of their difference. But we have to resist those viewpoints that would like to entomb the discoursing, organically articulated human being inside Hübner's gear-and-button automaton.

In order to clear away these sorts of questions and to firmly establish the difference between unconsciously occurring absolute self-production and the conscious reconstruction of an organic structure, we offer the following parallels.

According to a notice in *Das neue Blatt*, Götz of Berlichingen's famous iron hand, which a blacksmith at Jagsthausen made for him, is still on view at the castle there. "Each finger of the hand can be opened and straightened by pressing a button to engage a spring mechanism, just like on a gunlock; when closed, the fingers clamp tightly around the sword. The thumb and carpus are fitted with a unique mechanism."[17] The mechanic most likely had a human model in front of him while he worked. And, in addition to a living hand, he probably had a well-prepared skeleton of the hand to help him in constructing the joint, which he did meticulously, connecting piece to piece, replicating the forms and relations he observed in the model. The mechanic then assembled the individual parts, with impressive skill for the time, so as to provide the iron structure with mechanisms of compression and elasticity sufficient to substitute for the hand in time of need (Figures 20 and 21). All so-called "artificial limbs" belong in this category. What else are they but *mechanical frames* [*Gestelle*]?[18] Seen from the outside, they often resemble their models almost to the point of deception. But what does this actually achieve? A supplement for a missing limb, though it may serve a specific individual purpose, can have nothing to do with the maintenance of general human welfare.

Fig. 20.

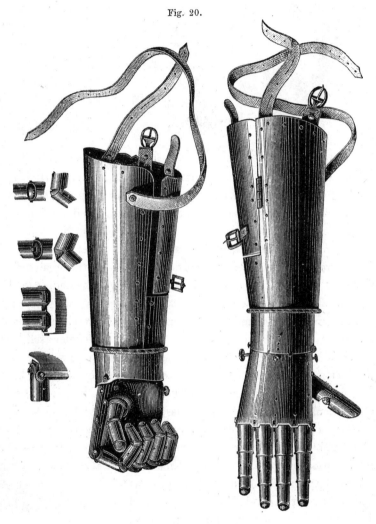

Figure 20. Iron hand of the knight Götz of Berlichingen.

The iron hammer and the iron hand appear fundamentally different from one another. The former is the effluent of an unabbreviated vital activity, while the latter is an anxiously faithful replica, painstakingly

Fig. 21.

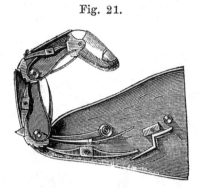

Figure 21. One finger of the iron hand.

re-created. The former enhances natural power and strength, the latter is a shabby refuge in face of weakness. The former is reproductively correlated with a whole series of tools, the latter an isolated object used to mask a deformity. The former takes a principal part in the creation of cultural means, the latter, while is it valuable to its owner, is for everyone else a mere curiosity. How imposing the analogy is—of the organic structure as the "tool begetting tool"[19]—in comparison with the sterile gadgetry of artificial limbs and the uncanny display of waxwork automata. The hand hammer is a hand metamorphosized, while the iron hand is merely its frame. Someone has to manufacture the latter; the former itself helps forge new hammers, erect entire hammer mills, and make world history.

In fact, it is as if the hand tool contained an organic endowment, in the idea of *handiness*; as if it were precisely this handiness—that is, the feeling of commensurateness the workman has with his limbs while he works—that enhances the intimacy he feels with his favorite tool, his sense of its indispensability. This is how we should understand the now proverbial "gilded ground of the handicrafts."[20] Then we can read in a rather surprising new light the explanation Caspari gives as to why primitive human beings, though they made fetishes of any number of trivial things, never idolized their tools, no matter how highly they prized them: it is Caspari's opinion that this fetishistic reverence was inhibited by the familiarity "that grew too intimately intwined with those objects—like the human body with its limbs—that he certainly should have prized above all."[21]

So long as the hammer—to stick with the standard example—is not recognized as the agent of industrial and plastic arts activities, in light of its endowment from the archetypal image, then we are still a long way off from the correct valuation of the primitive tool as the founder of human destiny. When one pauses to properly consider the hammer and to envision objectively what and how many marvelous works have been accomplished by means of it since time immemorial, "the ordinary thing" impresses and enthralls in the long run in quite another way than do those dancing, clattering artificial clockwork dolls that quickly turn from objects of curiosity into objects of tedium.

The products of organic projection were initially, as we have seen, of crude sensuous simplicity, serving only to facilitate difficult manual work. Gradually, we watched them assume, in multifarious configurations and sophisticated designs, the form of scientific tools and apparatuses to serve the ends of spiritual activities. At this point, the formal after-image begins to give way to the representation of inner relations and proportions. How many different spiritual elements the piano key displays, by comparison with the sledgehammer! Yet both are products of the need to give form, and both recall certain organic forms and movements. The piano key, because indicative of the ideal realm, is an instrument of art, regardless of the fact that there is only one way to use it; whereas the hammer is simply a hand tool, albeit the factotum of handwork and the condition of possibility for all of art's tools.

We might consider the example of the magnetic needle—as the mechanical expression of the polarities active within the organism—as tentatively alluding to the sort of forms the projection of ideal relations will have to pass through in order to become, ultimately, a total image of the bodily organism as an organism of spirit. Alexander von Humboldt referred to tools of the ideal type as *new organs*, and we find confirmed in this what we are referring to as organ projection: "The creation of new organs (instruments of observation) increases the human being's spiritual and also often his physical powers. Closed electric current conveys thought and volition to the furthest distances faster than light. Once the forces operating silently and imperceptibly in elementary nature and in the delicate cells of organic tissues are recognized, deployed, awakened to higher activity, and entered into the incalculable sequence of means, the human being will be brought nearer to mastery over individual domains of nature and to the living awareness of the universe as a whole."[22]

CHAPTER 6

The Inner Architecture
of the Bones

We are now entering that domain of organic life that might be considered, in a literal sense, as its *ossification*. Something remarkable is taking place in contemporary research: once pronounced dead, the skeleton is coming to life again. A few brief reports notwithstanding, this circumstance has scarcely impressed itself on our general awareness. Given the corroborating value these reports have for organ projection, it is more than appropriate to elaborate them here.

The situation—important in and of itself for physiology in general, and especially so for our purpose—is as follows:

It is a well-known fact that certain architectural rules are being applied to the design and construction of iron bridges, notably railway bridges—rules for which physiology and mathematics have discovered a surprising and unsuspected prototypal image in the structure of bone substance in animal bodies.

The vanguard of these discoveries—a scientific triumvirate whose research has converged on a common problem and reached a consensus seldom encountered—have expedited the disclosure of a great truth, releasing finding upon finding in unusually rapid succession shortly after having discerned the first clues.

Should the rationale for organ projection, so far as we have developed it, meet with any misgivings, the results arrived at by means of an inductive method—fostered not only by an opportune coincidence of circumstances but also by the brilliant approach to the topic—should effectively silence any doubt. In this case, I cannot forgo a description of how the events unfolded—though not for the sake of recalling details of general human interest, but because, as we know, the *course of events* of a thing bears on our *judgment* of the same.

The scholars involved include Karl Culmann, professor of mathematics at the Swiss Federal Institute of Technology in Zurich; Georg Hermann von Meyer, professor of anatomy at Zurich; and Julius Wolff, medical practitioner and lecturer at the Humboldt University in Berlin. It is a moot question as to which of these three scholars can claim a greater share of the fame when viewed from the side of the spirit where these sort of metrics have no purchase.

Culmann is the author of a multivolume work on *Graphic Statics* that appeared in its entirety in 1866 (the first half having appeared in 1864).[1] Meyer published a paper in 1867, in *Reichert and Du Bois-Reymond's Archive*, titled "On the Architecture of Cancellous Bone."[2] And these prompted Wolff to publish in 1870, in Virchow's *Archive for Pathological Anatomy and Physiology*, his paper "On the Inner Architecture of the Bones and Its Importance for Bone Growth."[3] Later, in 1873, Meyer reproduced his paper "On the Architecture of Cancellous Bone" in a new work on *The Statics and Mechanics of the Human Skeleton,* connecting this with the relevant scholarship, including in fact several remarks directly referencing Wolff.[4] It might also be noted that the journal *Der Naturforscher* included a brief discussion of the topic under the heading "The Architecture of the Bones" and, in close connection with this, a notice on "Bridge Engineering."[5] Otherwise, it appears on the whole that little attention has been paid to these scientific events. Meyer does mention Culmann's proofs that the architectonic arrangement of certain bones agrees with the theoretical lines of graphic statics, though he does not elaborate on this, remarking more generally on the correlation between the architecture of cancellous bone and the bones' static and mechanical proportions (Figure 22).

In the interest of clarity, we will quote liberally from Wolff's account of how the discovery unfolded. He describes as *"an extraordinary good fortune for the sciences* that Professor Culmann of the Society of Natural Scientific Research at Zurich had the opportunity to examine Meyer's specimen. Had this not been the case, it is possible that Meyer's valuable discovery might have gone on being viewed as one of nature's elegant though meaningless pranks.

Fig. 22.

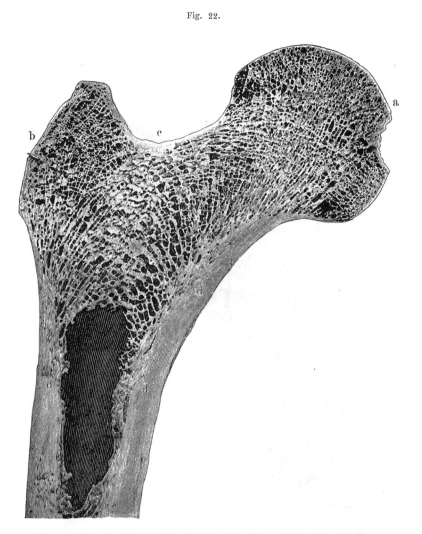

Figure 22. Veneer sheetlike frontal longitudinal section of the upper end of the right femur.
a the femoral head, *b* the greater trochanter, *c* pit of the greater trochanter, *d* the compact structure.
Between *a, b, c, d* the spongiose structure.

"Culmann remarked immediately upon seeing the specimen that the cancellous trabeculae in many places in the human body are constructed in exactly the same lines as those that mathematicians developed in graphic statics on bodies having shapes similar to the bones and which are subject to similar forces. He drew a crane (i.e., a bent beam meant to lift or bear loads) in the shape of the upper end of the human proximal femur, to which he assigned a load equivalent to the conditions in the human body. He then had his students draw so-called *tensile* and *compressive lines*

Fig. 23.

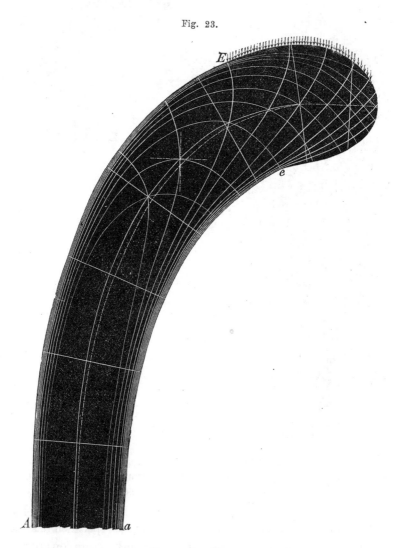

Figure 23. Bone-like crane.
Progression of the theoretical tensile and compressive lines, *A–E* tensile lines, *a–e* compressive lines.
The greater trochanter is thought to be broken off.

into this crane under his supervision (Figure 23). The result was astonishing! It turned out that the lines corresponded at each point with those that nature had realized at the upper end of the proximal femur in the directions of the bone trabeculae."

During a stay in Zurich in October 1869, Wolff examined Meyer's specimen and became convinced that the cancellous bone structure is in fact not a disorganized meshwork of trabeculae and cavities, as it had been believed to be, but that it is better characterized as having a well-motivated architecture. From that point on, Wolff concentrated intensively on the bones' architectonic proprotions, meanwhile maintaining a scientific exchange with Culmann. In all his discussions, he kept to one specific example, describing only the proportions in the proximal end of the human femur.

While Meyer had prepared his specimen by sawing the bone in half longitudinally, Wolff had occasion in Berlin to use a machine originally intended for cutting paper-thin ivory platings for women's accessories. The bone, affixed onto a little wooden block, was then slowly drawn in a horizontal direction through a fine, foliate saw moving up and down with tremendous speed. A single bone yielded many so-called veneer sheets; after having been cleaned and made transparent and laid over a base of black silk to produce a strong contrast, these could be photographed in clear, exquisite detail. Wolff had included photographs like these in his paper, and the juxtaposition of Culmann's drawing of the femur specimen with his own schematic reproductions showed that "the correspondence between the theoretic tensile and compressive lines on the bone-shaped crane and the structure of the trabeculae was *perfect.*"

Wolff further showed "that load on the flexion resistance stresses both the femur and the crane, such that all the particles on the side of the greater trochanter are pulled apart and all the particles on the opposite, inner side (adductor side) are compressed. This said, the former should be termed the tensile side, the latter the compression side" (Figure 24).

"However, tension and compression are not the only effects of an exterior force putting stress on the flexion resistance of a body. The particles of any cross section in a material attempting to resist flexion tend to shift their position with respect to the particles of the neighboring

Fig. 24.

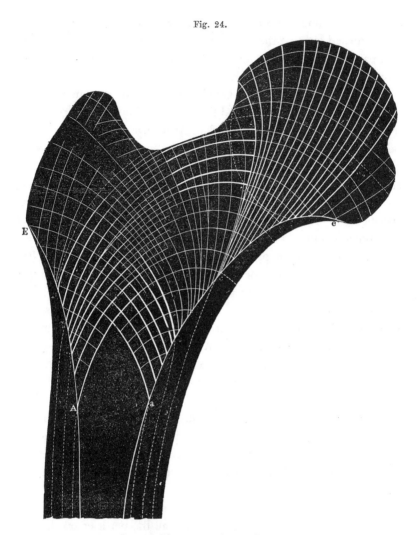

Figure 24. Schematic reproduction of Figure 22.
A–E tensile lines, *a–e* compressive lines.

cross section, and the particles of any longitudinal section tend to shift their position with respect to the particles of the neighboring longitudinal section. The force by which this happens is called the *thrust*, or *shearing force*, and it results in a *shear stress* in every section, which withstands the movement of two neighboring sections against each other."

Wolff fills nearly three pages of his paper with a demonstration of the above; from these, I have selected the following, in summary of the most significant findings: "According to Culmann, by eliminating shearing forces, a material is better able to withstand the tension and compression of a load and in this way is able to bear as great a load as if it were a solid mass. By the same method, the most practical bridge design is able to prevent shocks and vibrations while also keeping both the material expenditure and the cost of bridge girders to a minimum.

"The calculations discussed above are not merely theoretical, but have already been applied in practice. I should not fail to remark here that Pauly's bridge truss is based on the theory of tension and compression lines according to which the bone is constructed. It is confirmed that the so-called *compact substance* of the bone is actually pressurized cancellous bone, while its individual layers should be considered directly continuous with and supportive of the trabeculae of the cancellous portion. . . . One can distinguish clearly between a compact and a cancellous region of the bone, but the past distinction between two bone substances no longer holds."

Wolff then adds that *nature has constructed bone like the engineer constructs a bridge,* but that nature constructs both more completely and more splendidly than the latter ever could. This goes to show how determined Wolff is to recognize nature's power of demonstration everywhere—notably when he admits that *what impressed him from the outset about the proportions in the architecture of the bone was that here it seemed he was dealing with a matter of mathematically predetermined fact retrospectively confirmed in reality* (Figure 25).

In the final section of his paper, Wolff discusses what the inner architecture of the bone implies with respect to bone growth. Rejecting the theory of apposition of new bone substance to the already existing material, he overturns the old dogma of mechanical passivity (juxtaposition theory). From his own standpoint—the only one worthy of the organism—he asserts that bone growth is an organic (interstitial) metabolic renewal (intussusception), in this way *rightfully placing bone tissue among the active, living tissues.* Consequently, he is able to state quite explicitly "that here *we have a general law that holds for every region of the body, that bone possesses an architectonic structure answering to its use."*[6] From only the external form of the bone, he is able to infer without fail whether only its resistance to pressure will be stressed or its resistance to flexion

Fig. 25.

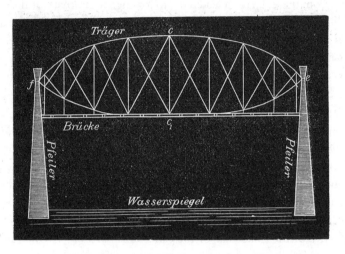

Figure 25. Schematic representation of Pauly's bridge truss.
f c e and *f c₁ e* are the two practically feasible tensile and compressive beams, the inner tensile and compressive lines are replaced with trusses. The bridge is suspended from the abutments in neutral axes at *e* and *f.*

as well and therefore which architecture should be found in the cancellous substance; and, conversely, how the architecture of the cancellous bone is indicative of the use and the load-bearing region of the bone. The practical investigations he has undertaken in this direction have made certainties of his initial conjectures.

"The wonders of the cancellous region of the bone" are now revealed. As Karl Rosenkranz remarks in his *Psychology*, hard bone *must* tell the truth.[7] But hard bone will have to speak many new and unimaginable things before it will be able to corroborate Goethe's assertion that "there is nothing in the skin that is not in the bones."[8] We have known for some time that the cranial bones help with hearing, since "their texture and shape make them excellent conductors of sound and therefore they seem to cooperate generally in the reception of sound waves."[9] There are even "animals that hear exclusively by means of bone conduction—fish, for instance."[10]

This research has satisfied the principal concern, insofar as it has cut a path along which a series of conclusions can be made. One may begin from any bone whatsoever, the form and composition of which first awake a representation of the complete skeleton; this representation then extends to the architecture of cancellous bone, further to the load-bearing region of the bone, to the musculature and organ functions, to the nutritional state, even to physiological processes, in this way forming a sequence of inseparable links in a chain!

It now remains for us to explain how the discovery of the architecture of cancellous bone, as a key support for organ projection, operates with respect to the double direction of organ projection.

This double direction may be briefly summarized as, on the one hand, the unconscious displacement outward of the bodily into its material after-image and, on the other, as the reverse application of the mechanism in order to understand the organism, effecting the greatest possible increase in conscious awareness of the unconsciously realized conjugation [*Ineinsfügung*] of an interior and an exterior.[11] And so we find represented in this analysis of the architecture of the bone primarily the one side—that is, the latter.

The spongelike structure of bone substance, the cancellous bone, has been in plain view all along. Primitive human beings knew it in the same way that modern anatomists do: by splitting the bone longitudinally with an ax to release the pith within. The meshwork composed of tiny beams and plates lay right there, whether or not it became the object of extensive research. And that was that.

The skeleton's general purpose once appeared obvious and in need of no further explanation. As long as the bone was not classified as organic, the means by which nature had achieved this purpose, however, could not be discovered. Its growth seemed, like the changes in inorganic bodies, a matter of enlargement through apposition of new bone substance to old. The skeleton was thought to be a sort of support structure that, like wood or stone, would outlast by centuries the decomposition undergone by the rest of the body. But, as the concept of the organism was gradually clarified and developed, and as the insight was consistently reinforced that the purpose of an organic part—a limb, in other words—is its sole determination (or, as Wolff put it: *the manner in which the bone is constructed is its only possible architecture*), one began to regard these so-called dead bones in a different light.

Against the doctrine of mechanical apposition, one sought evidence for organic growth from the cell. It turned out that the trabeculae were formed as muscle and nerve fibers are, their shape and distribution reciprocally determined in connection with the whole. Cancellous bone proved to branch out from compact bone, the latter in turn proving to be extremely condensed cancellous bone, the lines of which, tensed in particular directions, were suspected of being coordinated with the known purpose of the bone to withstand pressure and load.

Meanwhile, technology had been erecting mechanical apparatuses over canyons and bodies of water, the forms of which, derived through mathematical calculation, corresponded to the exact same purpose as that of the bone. Coincidence happened to bring the right men together in the right place and, as if scales had fallen from their eyes—for "we do not think," as Adolf Bastian writes, "it thinks in us"—a marvelous discovery "occurred," a new, dazzling light flared in the sciences!

It is the sophisticated *mechanistic disciplining of natural scientific research* that has enabled this and so many other great discoveries in the field of physiology. An existing mechanical apparatus serves the researcher—who is unaware that its form emerged through a process of organ projection, though not unaware of something within himself urging the comparison—as method and tool for the disclosure of the miracle of organic construction. Accordingly, Wolff was able to declare that *"nature has solved a mathematical problem*, so to speak, and provided a wonderful confirmation of the theory of tensile and compressive lines." *Thus the mechanism is the torch that illuminates the organism.* This is the standpoint of Wolff's discovery; this, as was mentioned above, which is one side of organ projection, is a manner of testing upon one's self.

But how is this even thinkable without assuming the other side? This other side—the actual task of our present reflections—has gone unremarked until now and seeks its solution in further factual evidence of the essential conformity between the projecting organ and its projected technical after-image.

Far from criticizing the detection and elucidation of organic conditions by means of prior experience with machines and apparatuses, one should rather attempt to satisfy the rigor of the inductive method. Tools, apparatuses, and machines exist; how they came to be is the object of a later understanding. No part of them escapes the scrutiny aimed directly at it; the eye can see and the hand can grasp the forces made to operate on the basis of mathematical calculation. *Physiological processes, on the other hand, cannot be immediately understood; rather, they must be grasped experimentally with the help of mechanical apparatuses.* The mechanisms that assist us in this most directly are those whose properties display an obvious affinity with organic fixtures. The value and significance of the means that enable our recognition of this affinity necessarily increase when they are successfully applied. It should therefore come as no surprise when experts, like Max Heinze for instance, observe: "think about the mechanical operation first, or at least *for all operative causes, look for an analogy in the mechanism.*" Nor does it come as a surprise that, in order to communicate new physiological findings, experts have to fall back on the parlance of mechanics.[12]

The way in which science provides itself with insights is naturally the one best suited to its making itself understood exoterically.

And so, when physiology depicts the living using the forms of intuition it has acquired through use of and reflection upon the mechanism, these depictions should command all the more respect from us—for without grasping at facts and comprehending them mechanistically, we would never have made many of the most important and beneficial discoveries. Of course, it takes time for teleological truths to shed the last of the moult from their origins; even today astronomers use Ptolemaic instead of Copernican expressions when they want to be widely understood. It follows that the mechanistic conception of the living consists in just this way of speaking, which, as a sensuous depiction, should be regarded as a tool for understanding ideal relations.

The rapport existing between the mechanical apparatus and a particular organic structure is predestined. In this way, the eye and the magnifying glass may be found one within the other, so too the ear and the vibrating string, the heart and the pumping station, the larynx and the organ's pipe, the femur and the bridge girder, the hand and the hand tool. Once recognized, this rapport will be technically exploited in the most diverse ways in its conscious transmission beyond the original relation. Thus the diverging lens is perfected in the ophthalmoscope, and thereafter other organs too—like the larynx, ear, mouth, nose, and so on will be submitted to lights and mirrors with the same success. Unconscious, instinctive finding and conscious, deliberate seeking and invention increasingly dissolve into one another. Even in cases where we are certain that an extremely elaborate apparatus can only have been the product of expertly trained deliberation, further reflection will always lead us to the realization that the elements of the composition, fetched from the sphere of the unconscious, point back each and every time to the original—that is, the unconscious—work of the hand. *However brightly the conscious creative works of technology may shine before our eyes, we know this light to be reflected from out of the depths of the unconscious, indeed to be no more than the consciousness first unbound in primitive tools!*

And so we find ourselves pressed further and further along the trodden path of *actual empiricism*, to the ineluctable conclusion *ab interiori* that everthing emanating from the human being is in fact his own nature—human nature—which he disperses throughout the world and which is displayed before him as the system of human needs. Authentic consciousness then comes about as a result of his gathering together *ab exteriori* his own dispersed nature to rediscover in bodily being the unifying point of departure.

The system of needs—the substance transfigured via organ projection into tools and equipment and impregnated with something spirit-like as well as the changes these have effected over the face of the earth—is the outside world, which has the following advantage over the other, natural world in which the animal is also situated, though it will always be alien to the animal: in this outside world the human being becomes self-conscious, truly rediscovering himself in it and learning to know and to comprehend himself.

For if the human being did not possess the original disposition to create in his tools artificial organs with which to comprehend the stimuli affecting him from without, he would never behave toward nature, toward the cosmos any differently than does the animal, for whom the world remains forever uncomprehended. To be sure, nature is the first condition of existence for every organism; but, when compared with the outside world created by the human artifactive drive, nature is not the first instance of human perfection.

The expression "principle of actualism" was first introduced into natural scientific research by Justus Roth and Emil du Bois-Reymond. Paul Samt defined it more precisely as the "principle of empirical actualism"—or, "actual empiricism," which is how we referred to it above. The expression, in spite of these changes, retains its demonstrative force. "The principle demands," writes Samt, "that actual empiricism is always the premise, and that what is actual is assumed as fact *even for the past and the future*, until the opposite is proven."[13]

We will use the full weight of this method of demonstration for organ projection. Since we have already established the efficacy of organ projection in the past and the present, we will have to prove it effective for the future as well. It is up to us, by carrying on enlisting more facts, to allow this efficacy to manifest as one that is vouchsafed to it.

CHAPTER 7

Steam Engines and Rail Lines

W e are now leaving the domain of those works of technology that, though universal and widely available in countless variety, we have to this point considered in isolation. We turn our reflections now to the massive cultural means that, like the railroads and telegraphs that span the entire globe and link all the world's parts, far exceed the term "apparatus" and appear to us as *systems*. But before we can discuss the railroad as a system, we need to understand a single factor of it, the *steam engine*, since all that holds true for the steam engine also holds true for this one particular form of its application—the locomotive.

In the rotunda of large-scale industry, the steam engine, as a machine in the superlative sense, is *the machine of machines,* in much the same way that the hand tool is the tool of all other tools in the domain of individual mechanical forms. Once industry, conceived in and gradually evolved out of handicraft, had expanded sufficiently to draw on wind and water power, the human being found that he was capable of handling enormous quantities of matter and, moreover, that he was able, through regulation and exploitation of natural forces, to reserve a significant portion of his own physical strength that he had previously had to expend directly in the process of work. But the power of wind and water is intermittent, and the use of it—seafaring is no exception—depends on both time and place. The human being is subject to changes in the weather and season and to nature's dominion generally, and even though he may attempt to contain nature for his own purposes by means of dams, floodgates, and gear mechanisms, still nature by and large makes no attempt to accommodate him.

By now it has been one hundred years since James Watt perfected the steam engine. The ancient elements—a tight phalanx of earth, water, air, and fire—obeyed its summons. A comprehensive new motor had been achieved, and this marvelous invention began its tour around the world. This is where large-scale industry begins.

With the mastery of steam emerges an engine power of universal accommodation as well as a new concept of work, and this is simple to explain. Mass work requires masses of workers. In a certain respect, the regional concentration of workers is perceived as a sequestration, and "the worker"—feeling constantly a being-for-itself and a being-among-themselves—soon comes to believe it must be privileged, vis-à-vis the other professions, as an estate of its own.[1]

We do not have the space here to further elaborate this phenomenon. In fact, we mention it only because, bound as tightly as it is with the entire cultural inventory, the worker phenomenon bears witness to the world-historical significance of the first machines to have been built on the analogy of the concept of the *conservation of force*. The full impact of this on the future is at present beyond the realm of measure.

Known, admired, and in use the world over, the steam engine is truly a "universal all-purpose machine." It facilitates all human activities, in the home and yard, in forest and field, on water and land; it functions as draft and pack animal, helps lay cable and print books; and, given the universality of its capabilities, it is particularly qualified to serve as the sensuous illustration of the principle of the conservation of force. The steam engine is frequently evaluated in comparison with the bodily organism. Among others, Otto Liebmann writes that "indeed, we find many remarkable analogies. Both display a complex system of interrelated, articulated parts whose movements may be integrated; both are competent to perform certain kinds of mechanical work. The locomotive, like the animal, needs to be fed in order to convert the heat generated through chemical oxidation processes into a system of movements. Both secrete waste, products of combustion, in more than one state of matter. There is wear and tear and exhaustion of machine parts as there is with organs. In both we find the interruption of all functions and death either when the input of fuel or nutrients ceases or when an essential part of the machine or the organ is destroyed."[2]

Helmholtz elaborates at length on this comparison in his lecture "On the Interaction of Natural Forces": "Now then, how does the organic being move and work? To the last century's builders of automata, humans and animals appeared like clockwork that never needed winding, that created its own motive power *ex nihilo*; they did not yet know to make the connection between nutrients consumed and the generation of power. *But since we have learned from the steam engine something about the origin of this power of work,* we have to ask: does it work the same for human beings? In fact, the continuation of life is bound to the continuous intake of nutrients. Nutrients are combustible substances that, once completely digested and taken up into the blood, are actually submitted to a slow combustion process in the lungs and finally pass into the air in the same oxygen compounds that arise from the combustion of an open flame. . . . The animal body therefore does not differ from the steam engine by the manner in which it produces heat and power but by the purposes to and the manner in which it applies the power thus generated."[3]

J. Robert Mayer puts it very similarly in his lecture "On Nutrition." After showing how animal nourishment differs from that of plant life, he goes on to say: "The animal further distinguishes itself from plants essentially through its ability to generate voluntary movements. But the material required to perform this mechanical work originates in the plant world, though even in the world of plants we are dealing with an antecedent energy source in the sun. Therefore, the animal actually converts what had once been sunlight into heat and motion. In this respect—I say, in *this* respect—the animal organism, given the limitless variety of ways in which the whole of it may be broken down into analyzable parts, may be compared with a steam engine. The steam engine too consumes the sunlight stored in the world of plants in order to operate, to generate work—and produce heat; and we cannot help using such a comparison now and again, with respect to animal nutrition as well as human—which, with regard to its body, has very much in common with the animal."[4]

With recourse to such qualified experts as Mayer, who discovered the mechanical equivalent of heat, and Helmholtz, who developed this theory into the law of the conservation of force, we hardly need cite any further evidence, since between the two of them what is pertinent to the *comparison of the standard machine with the standard prototypal image of all machinery* is more than sufficiently covered. But the comparison itself is only pertinent when it is complete, and it is complete when, apart from detailing all the points of agreement, emphasis is placed upon the general, characteristic difference that alone affords meaning and significance to the agreement that has been discovered. In this respect, one might follow with particular interest how resolutely these experts preserve the concept of the organic from obfuscation through admixture of the mechanical. Mayer is compelled to add expressly, that while comparison is founded on the detection of similarities, similarities are nevertheless a long way from identity. "The animal is by no means a mere machine; it even stands high above the plant, since it has a will." Helmholtz draws the distinction between machine operations and human work even more explicitly: "When speaking of the work of machines and natural forces, we must naturally be careful to exempt the activity of human intelligence from our comparisons. What appears in the work of machines as intelligent action naturally belongs to the intellect of he who built it and cannot be credited to the machine as work. . . . The concept of work has obviously been transferred to machines, given that their performance is often compared with that of animals and human beings, whose performance the machine was meant to replace. . . . The clock's gear mechanism therefore does not generate any work power that is not first imparted to it, but steadily allocates over a period of time that which was already imparted."

No less pertinent is Liebmann's comparison of the machine with the human organism, in which he puts the obvious similarities aside to emphasize the difference: "But! But! The machine is an extrinsically and deliberately made artifact, while the organism is grown *ex ovo* according to a hidden, immanent law. The *hegemonikon* of the machine does not belong to it, does not reside in it; the stoker and engine driver sit on top of it and direct it, like the rider his horse. The ἡγεμονικόν of the living organism—*intelligence* and *will*—belongs to it, rests within it, originates with it, forms its integral constituent parts. And—without even taking its functions into account—the machine parts are there permanently; its constituent parts remain identical with themselves, until the point that external repair becomes necessary. But the organs of the organism remain self-identical in form only; their substance changes constantly, as they regenerate and repair themselves."

Explanations like the above help protect against the degrading mechanistic worldview that imagines the becoming-machine of human beings and the becoming-human of machines. Helmholtz's expression, *that the concept of work for machines was taken from a comparison with human beings*, immediately implies that the machine itself, if it is meant to substitute for human work, will be constructed to correspond with the organism whose work it is supposed to perform. The machine's ability to operate—or rather, its usability—is immediately correlated both with the human being that uses it and with the purpose the organ's own activity was intended to achieve prior to any mechanical support.

The individual tool manifests more or less recognizably both the operative capacity and the form of the organ at the same time. With complex machinery, the former stands out much more prominently, while by contrast the latter tends to recede. The form of the steam engine as a whole and that of the human body have little or nothing in common in appearance, albeit particular machine parts may resemble individual organs. Many machine parts, originally isolated tools, are externally assembled in the steam engine to produce an overall mechanical effect, while the limbs of the animal link up inwardly, from the simplest to the most complex organic living entity—the human being.

In this way, the *theory of organic development* corresponds with a *practice of mechanical perfectibility*—ascending from the primitive stone hammer through all tools, apparatuses, and machines of simple construction, to that complex mechanism in which the *model machine* may be recognized. We say *model machine* because science perceives its value as a tool and as a kind of physical apparatus through which to understand the interaction of natural forces as well as the life processes in the organism. "The invention of the locomotive engine may be said to have begun when the first men learned how to make a fire and keep it alive with fuel; another early step was the contriving of a wheel; command was won, by degrees, of the other mechanical powers, at first in their simplest, then in their more complicated forms and applications; the metals were discovered, and the means of reducing and working them one after another devised, and improved and perfected by long accumulated experience; various motive powers were noted and reduced to the service of men; to the list of such, it was at length seen that steam might be added, and, after many vain trials, this too was brought to subjection—and thus the work was at length carried so far forward that the single step, or the few steps, which remained to be taken, were within the power of an individual mind."⁵

The greastest inventions turn out to be products of an ongoing process of self-finding, whose aim the human being is unconscious of at first. The preparators positioned along this protracted world-historical boulevard of inventions leading to the steam engine were sufficiently conscious of their present individual purposes, though unconscious of the great cultural idea their work hastened to manifest in the locomotive. Just as finding and intentionally devising persist in reciprocity, *so do the conscious and the unconscious incessantly displace and work through one another.* But prior to the moment that the idea is actualized, the restlessness of conscious searching has the upper hand. After a succession of individual inventions have stripped away enough layers, one begins to discern the idea through the few translucent layers remaining; at last, the idea occurs to an individual who perseveres in his research and has the courage to perceive. James Watt perceived clearly and explicitly what he was looking for, and therefore he succeeded, when the time was ripe, in accomplishing the desired invention on the ground that all prior experiments had unconsciously prepared for him—though even Watt could not have foreseen the next, higher phase to which Stephenson would take his invention.

Though existing simultaneously, for a long time the steam engine and rail lines were alien to one another. Then Stephenson provided the steam engine with steady and reliable forward motion and, by submitting rail lines to the *locomotive*, he created railroads. So long as rail lines and the steam engine existed independently of one another, the rails themselves represented only a slight improvement in conditions for the familiar animal-powered mining trains, while the steam engine remained little more than a wind- and water-power substitute that could be erected anywhere. But, united to form the *railroad network,* and further expanded in the *steamship lines* that crisscrossed rivers and oceans, rail lines and steam engines have become our contemporary medium of universal communication, the mediator of the pervasive human presence the world over.

Rail lines united with steamship lines to form a closed system, the network of transport arteries through which circulate humanity's means of subsistence—a likeness of the blood vessel network in the organism. From our perspective, it certainly would seem strange if, in its depiction of the circulation of blood, science had failed to seize on the available analogy between the organic process and the mechanism by which the life requirements of human beings are circulated. Time and again we come across remarks indicating that in this case it is nearly as impossible to avoid recognizing organ projection as it is to avoid the by now habitual comparison made between the nervous system and the electrical telegraph.

In light of this, we can appreciate the comparison Heinrich Oidtmann had occasion to make when describing blood circulation in a recent public lecture. According to the report in the *Kölnische Zeitung*, "Oidtmann managed to provide the audience with a clear idea of the complicated processes of blood circulation by invoking an image of a railroad network, with a set of double rails, with through-rails and connecting rails, with stops and in- and out-bound lines." It should be self-evident that what is meant by "image" here is not the allegorical figure that we have mentioned at several points already. That is, the "image" is not a pure simile that puts at one's disposal an apparently unlimited number of possible comparisons; rather, it is the concrete image and reflection of the projection, which exists uniquely. In a talk concerning steps that needed to be taken in order to improve the transportation system, Franz Perrot first gives a detailed account of existing transportation facilities on land and water before concluding with the following: "It is a self-evident matter of fact that transportation's several branches do not exist independently of one another but instead interlock, reciprocally condition each other, and together form an interrelated whole that is approximately for the state what blood circulation is for the human body."[6] Likewise, Max Perls puts "the great railway lines" on a level with "the communication system of the blood vessels."[7]

So much, then, for the steam engine, and for rail lines as the primary condition for its locomotion—a situation that we will encounter again later on in the context of other functions of the state.

What also appears so magnificent here is the prevalence of the unconscious in the creation of such a powerful lever of industry. Certainly, the idea could not have been far from the minds of James Watt and Robert Stephenson, that they might allow their own bodies to stipulate for them the law and the norm for the mechanical structure of their machines. That being said, the coordination between them is so exact that this affinity between the organic prototypal image and its mechanical after-image serves as a considerable source of evidence for even the most respected representatives and interpreters of knowledge concerning human beings. Why else this constant appeal by the sciences to the fittings of mechanical apparatuses and their recourse to the vocabulary of mechanics? But let us not forget that the unconscious, though it may recede when it comes to executing the particulars of technical engineering, is felt to be all the more operative when we index the organ's disposition to activity to the mechanical construction of forms. Since, as the steam engine so clearly illustrates, progress toward higher mechanics consists less in the unconscious reproduction of organic forms than it does in the projection of their *functional* image—that is, the image of the living and of spirit operating as organism.

What instills in us such great admiration for the steam engine is not any of its technical particulars—not, for instance, the reproduction of an organic articulated joint using rotating metal surfaces lubricated with oil; not the screws, arms, hammers, and pistons. Rather, it is the feeding of the machine, the conversion of combustible materials into heat and motion—in short, the curiously daemonic appearance of a capacity for self-motivated work. Here we are reminded of the higher provenance of the machine in the human being, whose hand built the monstrosity and set it free to compete with storm and wind and waves. Every scrutinizing glance cast upon it contributes to establishing the truth of Feuerbach's proposition that *the object of any subject is nothing else than the subject's own nature, taken objectively.*[8]

CHAPTER 8

The Electromagnetic Telegraph

When one wishes to indicate the immediate conjunction of one process with another, one might use the expression "to follow on the heels of"; and so, one could say that following on the heels—or rather, the rails—of the railway was the *electrical telegraph*.

The comparison between the electrical telegraph and the *nervous system* is self-evident. It has entered general use to describe the behavior of electrical current in the organism. In fact, our everyday representations of nerves and electrical wires coincide so well that one can justifiably claim there is no other mechanical apparatus that agrees this accurately with its organic prototypal image. Moreover, there is no other organ whose character is displayed as clearly in the construction that has been unconsciously formed in its image than the nerve cord is in the telegraph cable. Here organ projection celebrates a great triumph. Its principal requirement being: production taking place unconsciously on the model of the organic; followed by an encounter in which the original and the copy find themselves, compelled by the logic of analogy; and finally, revealing itself like a light switched on in the conscious mind, the agreement between the organ and the artifactual tool, with the greatest conceivable degree of identity. These moments in the process of organ projection emerge most clearly in connection with the telegraph system, and at this point we will refer ourselves to one who is most competent to speak to the issue.

Rudolf Virchow says in his lecture "On the Spinal Cord": "If you were to cut straight across one such fiber (nerve cord), you would find individual bundles bulging out from the cut surface in the form of whitish protuberances. You would then have captured an image that *exactly* corresponds in miniature to a segment of submarine telegraph cable. Just as you might remove the insulating layers to strip the individual wire, you can also defibrate individual bundles of nerve fibers from the nerve sheath and further separate the individual nerve fibers from these bundles. *In fact, the proportions are exactly the same: the nerves are the cable installations of the animal body, just as you might call telegraph cables the nerves of humankind.*"[1]

What more could be desired than what is already contained in this remark? Every all-too-cautious "quasi" and "as it were" is extinguished by this categorical "in fact," recedes before an explanation that permits neither caveat nor equivocal interpretation. The nerves *are* the cable installations of the animal body; telegraph cables *are* the nerves of humankind! And, to this we will add, *they must be,* because the defining characteristic of organ projection is that it proceeds unconsciously. Or did the men who first succeeded in sending messages across vast distances by means of electric current consciously intend, even before the first experiments in nerve dissection were conducted, to make an exact artificial reconstruction of the body's own nervous system and to lay this branching electrical framework over the entire earth (Figures 26 and 27)?

Elaborating in greater detail how what we know of telegraph cables is reiterated in nervous activity, Virchow continues: "If we knew absolutely nothing about the changes caused by excitement of the nerves, we would know nothing about nervous current, nor would the similarities with telegraph installations occur to us. But we do know, because of du Bois-Reymond's investigations, that in fact nervous current is an electrical current, and therefore we can easily say that the entire installation and activity of the human locomotor system can be placed in parallel with the layout and action of the telegraph.

"In and of itself, the motor nerve has no other property than to act as the bearer of nervous current, as it moves away from the spinal cord to the muscles—that is, centrifugally—and, having reached the muscle, there to induce self-movement. The current, as such, is not at all visible, *just as the current in the telegraph wire is not.* The active nerve looks just like the inactive nerve; it changes neither place nor form.

"The sensory nerves are distinguished in their *peripheral* course, in that they are not connected with any other parts in particular. Even the sensory nerve fibers go on ramifying, but they do so between the tissues

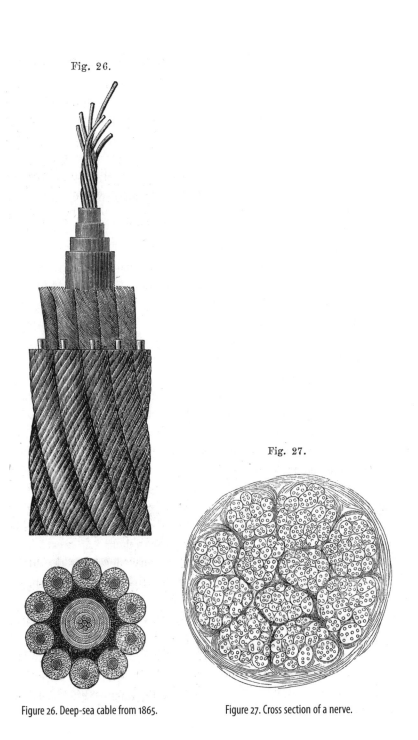

Fig. 26.

Fig. 27.

Figure 26. Deep-sea cable from 1865.

Figure 27. Cross section of a nerve.

and the majority of them terminate in stand-alone nerve endings. . . . The excitation of these termini is propagated on the nerve fibers connected with them, and a centripetal current results that is conveyed through the peripheral nerves to the spinal cord.

"Since we know that the strands of white matter in the spinal cord connect up with the brain, indeed merge into it, it is apparent that here we have before us another conduit that communicates with the brain and thereby with the peripheral nerves. Now, since the brain is the seat of volition and consciousness, the spinal cord thus forms the mediating link between the brain and nearly all other parts of the body, correlating both voluntary motion and conscious sensation."

Even if the above were the only evidence available to us concerning the "perfect" parallelism between nervous and telegraphic conduction, the above passages would still more than suffice for what we wish to demonstrate. Nevertheless, it is advisable to consult the work of other experts, in part to explain the subject as thoroughly as possible, and in part to show that Virchow's view has been widely accepted as the most probable, generally speaking.

Carus writes that "among the purely physical apparatuses, we have one that corresponds with exceptional similarity to the concept of what is required of the nervous system and nervous life, and that is the *electromagnetic telegraph*. . . . *In this apparatus we find the perfect likeness of that mysterious formation, the nervous system*. . . . The example of the electrical telegraph will be especially illuminating here, for it becomes immediately apparent that the telegraph's wires would be entirely meaningless if they did not provide a pathway for the periodic galvanic current issuing from the galvanic apparatus connected with them." After explaining that conduction alone is represented in the nerve fibers, and that in the organs of the central nervous system, with the blood's cooperation, the agent of innervation is steadily discharging, Carus continues: "If one is therefore able to compare nerve fibers with the fibers in the conducting wires of the galvanic telegraph, then the body's cells, the larger of which are called *ganglion sphere cells, are the perfect equivalent of the galvanic battery,* as that which excites the current in the wires."[2] In this and other similar ways in a number of other passages, Carus makes use of the telegraphic analogon to describe nervous activity.

Mayer also likens the ganglion-free nerve strands of the brain and spinal cord that control the voluntary movements of the heart and intestinal tract to "telegraph wires." Likewise Czermak, with respect to involuntary heart movement: "Owing to ongoing nutritional processes, a state of excitation unique to each nerve substance arises that is then propagated as a motor impulse within the nerve fiber—like electric dispatches on a telegraphic conducting wire—to the fiber of the heart muscle and there induces a contraction."[3]

Wundt actually extends the comparison to the end organs of the nerves, the nerve fibers, and provides us a detailed description of nervous excitation that bears out the comparison perfectly: "The nerve fibers are often likened to an image of telegraph wires, which always conduct the same kind of electrical current, though this current may produce many different effects, depending on the apparatus connected with the end of the wire—for instance, the ringing of a bell, the explosion of a cartridge, the movement of a magnet, the display of light, and so on."[4]

At one point in his book *God in the Light of the Natural Sciences*, Philipp Spiller introduces the comparison between nerves and telegraph wires, though here we can read a slight reservation in his use of the phrase "as it were"; whereas, in another text, he is unceremonious about calling the projection by its proper name, when speaking of the nervous vibration that is propagated "through the cable of the optical nerve" to the atoms in the affected region of the brain. But: that all the information that we need in order to see is sent from the retina to the brain via the single optical nerve is no less marvelous than, or just as marvelous as, the fact that one telegraph cable linking two stations contains an entire bundle of conducting wires, each of which can, via the battery, stimulate an activity. Spiller continues: "If the external stimuli are intense enough, the dispatch arrives via the sensory nerves at the central station with the command 'response required.' And the telegraph operator stationed in the brain is, in the healthy organism, so wonderfully quick and meticulous that it need only touch certain keys of the signaling apparatus in order to send a response, via the motor nerves, to only those muscles to which the outside world has made an appeal or which it has otherwise stimulated, in order to perform some sort of operation by deploying the tension forces stored within the muscles."[5]

The invention of the electrical telegraph makes such a remarkable impression because the motive force that is conveyed through the wires is the very same that we know of in connection with innervation and therefore also with our will and sensations. This motive force equally serves thought and the dispatch form of thought. Paul Samt, in his previously cited text, provides us an elaborate example of how the language of electrical telegraphic technology is already operative in the description of psychic states.[6]

The laying of the first telegraph wires must have preceded any exact experimental knowledge of galvanic current. The analogy between the current-generating galvanic battery and the ganglion sphere cells in which, as mentioned above, the agent of nervous activity is steadily discharging presupposes an act of organ projection—an act that is also detectable in the agreement between telegraphic and nervous conduction along bidirectional pathways.

Galvani discovered as early as 1780 that contact between two unlike metals would produce an electrical current. In 1800, Volta built an apparatus that would resist electrical current. In 1819, Ørsted observed that galvanic current would deflect a magnetic needle. Immediately thereafter, Johann Schweigger invented the electric multiplier. Michael Faraday discovered electrical induction current in 1832. And finally, in 1837, Carl August von Steinheil produced the first working telegraphic apparatus. Though all press unremittingly on toward the same goal, each of these scientific events corresponds to a single step in a graduated sequence of mechanical apparatuses, each unmistakably displaying the hallmark of organ projection: the unconscious reproduction of an organic prototype.

To be sure, Galvani's original assumption that his discovery would have a physiological application, with respect to muscle and nerve activity, was not forfeited in the course of its subsequent development—though it was not until the work of du Bois-Reymond that positive indication of the electrical behavior of nerves and muscles finally became available.

Moritz Meyer writes: "Du Bois-Reymond was the first to succeed in demonstrating the presence of specific muscle and nerve currents, through deflection of the magnetic needle, aided by a very powerful multiplier; to standardize precisely the laws of muscle and nerve current and the changes these undergo through muscle and nerve action; and to regard the *frog-current* as the end result of all the many electrical currents, with their several vectors, that are present in the nerves, muscles, and other tissues."[7] Du Bois-Reymond's work, *Investigations into Animal Electricity*, is in this way a great scientific achievement, aiding the intuitively self-seeking human being in self-discovery.[8]

Since by now physiology has accepted the telegraph system as evidence for the electrical behavior of the nerves, in this organ projection is directly, tacitly affirmed. It is the factual guarantee provided by organ projection that alone justifies our reliance on mechanical apparatuses in helping us to orient ourselves in the domain of the organic. We therefore welcome the contribution of Alfred Dove in particular, which cuts so to the chase that just one small step remains before full agreement can be asserted. Dove writes: "Indeed, we understand the mechanism of nature only after we have freely invented it after the fact—as we understand the eye after the camera; as we understand the nerves after having constructed the telegraph." We need not explain that the "mechanism of nature" here means the bodily organism. But is this free "invention after the fact" the same imperative *unconscious finding* that we know from organ projection?[9]

We are hard-pressed to find a better answer to this question—one that both takes natural science itself into account and upholds our commitment to the power of the unconscious as the defining characteristic of organ projection—than the answer provided by the example of Michael Faraday: "There is hardly another single human being who is responsible for such a far-reaching and consequential series of scientific discoveries than Faraday. Most of them came as a complete surprise to him, *as if he found them purely by instinct.* Faraday himself hardly knew how to put clearly into words the chain of ideas that led him to them."[10]

We have thus far had several occasions to remark on the conscious intention, meant to remedy some deficiency, that is bound up with organ projection's unconscious process of form-finding. Increasingly, in the course of making further improvements to the mechanical apparatus that was originally formed on the analogy of an organic prototype, conscious processes begin to encroach—until eventually, once a certain point of perfection has been attained, consciousness appears to be the single operative factor. Such would be the case, in particular, with improvements mounted on or fitted to finished machines and apparatuses, improvements that are purely the products of painstaking deliberation. But this is even more so the case with newer tools—whether they be directed at penetrating the far reaches of space, such as the spectroscope, or focused on exploring the microscopic parts of the living body, like the ophthalmoscope— the successful manufacture of which could only have been the direct result of tirelessly repeated experiment and great expenditure of acumen and skill. And yet: who is to say there is not also something else operating behind such clearly intentioned inventions of physics and physiology, something metaphysical, metaphysiological—in other words, the metaphysics of the unconscious?

We have seen the unconscious take part in the emergence of the tool's basic forms and in their nearest modifications. Is it supposed to cease participating once we apprehend that even the highest achievements of conscious artifaction ultimately, in their turn, unconsciously serve the progressive revelation of the entire human being as thought-entity [*Gedankenwesen*]?[11]

Our reflections are approaching the point at which the concept of the tool begins to expand far beyond the raw materials of which the mechanism is made, beyond the frame that ordinarily accounts for its content. It now extends even to sensuously intangible formations. In the end, it will be sublimated into the concept of means and tool in the highest and most general sense, its material supplied directly from the workshop of the spirit.

The electrical telegraph paved the way toward this higher sphere of organ projection. The very same motive power that, on the one hand, is conveyed along the wire in the outside world and that, on the other, is involved with innervation in the body's interior in both cases serves to communicate thought—though, somehow, the difference between mechanical and organic communication remains constant and is not lost.

We will now summarize in a short overview all that has thus far been said with respect to mechanical apparatuses.

Having begun with an understanding of the tool and the organ in terms of language, we found that the essential characteristics of the hand—its status as the natural and therefore always already finished standard tool—not only contrived to produce the first artifactual tools but at the same time provided the prototype for them. The hand shaped the tools meant to support it and to increase its power and dexterity, according to its need for support, increased power, and dexterity. Because of this quality of handiness, these are the hand tools. In the hand—the extrinsic brain—tools establish culture, because they participate in the rapport existing between hand and brain. Eduard Reich puts it succinctly: "The most well-developed brain and the most well-developed prehensile tool come together in those species of human being that are capable of true civilization, and this reciprocal relation of both organs is the source of all ability, all knowledge, and all wisdom."[12]

In both aspects of the primitive tool—its purpose and its form—one finds a conscious intention to eliminate a momentary deficiency or, which is the same, to manufacture an advantage—even though configuration of this material for a specific purpose happens unconsciously.[13] This unconscious process—by which organic structures, activities, and relations act with respect to mechanical apparatuses as a prototypal image does to its after-image, and by which the mechanism acts as a means of disclosure and understanding of the organism—is recognized as the real essence of organ projection. Furthermore, we found that a myriad of things proceed from the hand's disposition to configure matter—things that the human being views as a second world external to him that is distinct from the world of natural objects. But the equipment that he exerts himself to make—his hand tools, above all—lie closer to and are more allied with him than are the things of nature, things he is able to overcome through knowledge and use of them beyond his animal pleasure.

Throughout the metamorphoses of the primitive hand tool into all sorts of equipment for housework, fieldwork, hunting, and war, its rudimentary endowment remains recognizable. In cases where it is less so, ethnography and the study of root words provide an equivalent fund of information on the beginnings of culture. Other than the hand and arm, we also considered the lower extremities, in part as organic proto-typal images, and in part on the basis of the kinematic chain of muscle movements that are transferred to tools and machines. After discussing the prominent limbs and their use in determining measure, number, and time, occasionally justifying the naturalization of mechanical terminology in the vocabulary of physiology, we extended our procedure to include the sense organs. The fields of optical and acoustical mechanics yielded enough sober facts to effectively silence any potential objection that here symbols and similes are at play.

In evidence of the growing pool of supporting facts available to us, we demonstrated the striking congruity of the vocal organs and the heart's activity with each of their respective mechanical after-images. Following this, the concept of the organism was again secured against inundation by the mechanical—albeit willingly admitting that the human being is initially able to represent his bodily self only through the exemplary use of mechanical apparatuses he has invented and laws he has discovered on his own.

The principle of actual empiricism could then be invoked for organ projection, in its full evidentiary force, on the basis of an extensive report detailing the wonderful discovery of the inner architecture of the bone, as recounted in a paper by Julius Wolff.

After this, the mechanistic worldview—which, in conceiving of the emergence of tools and machines ideally, is comparable to an intellectual tool—was allotted its rightful place, as the power that establishes an orientation toward the organic world and a power that aids it conceptually.

Up to this point, we were primarily concerned with accounting for individual objects; we then proceeded to consider the mechanical reconfiguration of the structures of the nervous and vascular systems that extend throughout the entire body. If mechanical engineering has already reached its peak in the steam engine, insofar as the steam engine is the genuine likeness of organic vitality, in relation both to the idea of the conservation of force and to locomotion, then, with the electrical telegraph— given that it communicates both written signs and thought; given also that it has shed its base materiality—we have approached as near as possible the domain of the transparent forms of spirit. This domain is elsewhere elaborated under the heading "Universal Telegraphics" in my own *Philosophical Geography*.[14]

Meanwhile, we will have to come to an understanding regarding organ projection's share in a theory of the unconscious by creating a context in which we can situate the points of contact that to this point we have dealt with in isolation. In a further chapter, then, it will be a question of rounding out theoretically our previous investigations into the emergence and development of artifacts, in light of a science of machinery recently founded on the basis of kinematic theory. Above all, among the traces of the unconscious in exactly this science, the human being becomes aware of the interrelation between his bodily existence and the world outside him, which was once a world within him.

Therefore, in allowing himself to be led by that affinity between prototype and after-image, in measuring against himself the outside world he has created, an ever increasing self-consciousness is realized in him.

CHAPTER 9

The Unconscious

Since we have arrived again at self-consciousness, via the bridge of the unconscious, it seems we would benefit from a more precise delimitation of the zone within which organ projection essentially participates in the concept of the unconscious—a concept lately expanded to near universality. This is not an easy matter, because, as we know, definitions of a single object all too often differ. At this point, we would be hard-pressed to find two thinkers with the exact same understanding of how the unconscious clearly emerges from the living. At present, a fairly vigorous debate over the *Philosophy of the Unconscious* is coursing throughout much of the literature.[1]

Everyone has his own unique standpoint. By this we do not mean only where one's feet meet the ground, that place that no two human beings can occupy at exactly the same time. What we mean, above all, is the standpoint of one's own thinking, according to which things appear differently to the minds of different human beings. No one thinks to deny that there is much that is operative within us of which we are unaware. All previous, sporadic thought and writing on the topic have for the first time been compiled and set firmly in a scientific context by the independent and exhaustive efforts of Carl Gustav Carus. He published this work in the form of a history of the soul's development, titled *Psyche*. Carus himself calls it "a work he had long cherished, much pondered over, and had to rethink time and again." It was for him "the dense fruit of many years of study, over the course of which I endeavored to lay out—always in genetic sequence, unfettered by didactic methods, and in the simplest manner possible—what those moments of clearest reflection had caused to mature in my mind, as the faithful yield of a deliberative intuition."²

The book, having come into being under the guarantee of such a strong "scientific conscience," deals in its three main sections with *the unconscious life of the psyche, the conscious life of the psyche*, and with *what is transient and what is eternal in the psyche's conscious and unconscious*.

The first section is concerned with the initial formative processes in the human organism and its primary organization, which is unconsciously regulated by the idea; with the unconscious in the process of individuation within the species; with what belongs to the field of the unconscious once a psyche has become conscious of itself; and with pathological states in unconscious pyschic life. The following section grapples with the emergence not only of the psyche in animals but also of the psyche and the spirit in children; with the interaction between unconscious and conscious psychic life; with the growth of psychic life; with the psyche's acquisition of personality and character; with the various emissions of psychic life, according to emotion, cognition, and will; and, prior to the relatively short though consequential third and final section, the second one closes with a discussion of psychic health and psychic illness.

This is the rich material of a work that should be considered foundational for all future studies of the unconscious. It is evident that, with very few exceptions, this sort of assessment of the *pysche*—one that is concerned with appropriately acknowledging all the work that came before it—had been missing from the literature in question, a lack which seems to have had a negative impact on "the conscious continuity of intellectual work" that Karl Friedrich Zöllner has called for.³ As the Arab proverb puts it: *The credit belongs to the founder, even if his successor outperforms him.*

The successor who is least liable to be accused of having ignored Carus's work on the psyche, and who has at the same time outperformed it, is the author of *The Philosophy of the Unconscious*, Eduard von Hartmann. To be sure, von Hartmann declares expressly in the chapter that deals with his predecessors' takes on the concept of the unconscious: "The concept of the unconscious has hardly any purchase in the new *natural science*. A notable exception to this is found in the work of the prominent physiologist Carus, whose *Psyche* and *Physis* are essentially concerned with studying the unconscious in its relation to bodily and spiritual life. It is left up to the reader to decide how far he succeeds in the attempt, and how much I can have borrowed from him in my own. I would only add, however, that here the concept of the unconscious is clearly presented in its purity, independently of that comparatively small consciousness." Moreover, the expression Carus uses to introduce the *psyche*—"The key to insight into the nature of conscious psychic life lies in the region of the unconscious"—appears as the epigraph to the second part of von Hartmann's book, "The Unconscious in the Human Spirit."

Were one to take on the challenge indirectly hinted at above and critically compare the two works, *The Philosophy of the Unconscious* would almost certainly win: it is seldom that we encounter fewer idle words and find so many potential misunderstandings neatly avoided. On those points where the two texts agree, *Psyche* is indeed a formidable ally to the work that succeeded it, though where *The Philosophy of the Unconscious* departs from it, we clearly see the advantage in its having done so.

For Carus, the unconscious is first of all the principle subtending the historical development of the human psyche. He resists the conventional label of psychology, for the discipline as it used to be treated had since outgrown its former parameters and begun to stream into the field of psychophysics. Von Hartmann, on the other hand, who expanded and raised the unconscious to the principle of our orientation in the world, locates in the unconscious that which had always already constituted the ground of all philosophical systems.

According to von Hartmann, the unconscious in Carus's *Psyche* is more or less narrowly conceived as one metaphysical factor in a single philosophical discipline. In *The Philosophy of the Unconscious*, as the limitless ground of all life, the unconscious is the inauguration of an entirely new worldview. One would not be mistaken in regarding this as a panentheism of the unconscious, the depiction of which—however it may emerge from the conflict between those who advocate for a universal *un*conscious and those advocating for a universal *proto*consciousness— can hardly diminish the great benefit in having essentially quelled the alternatingly dualistic and monistic attacks over the relation between the concepts of soul and spirit and in having enabled the deeply personal Self of the philosophers to appear as the only possible starting point for all of philosophy.

It is not uncommon for a terminology that has been in use for a very long time to occasionally give way to another—especially when a certain progress is impeded by the former's insistence on maintaining hackneyed tenets.

If, given the recently expanded limits of empirical research, the old norms no longer hold, then the situation is often righted when some aspect that had been previously cast aside as insignificant comes to the fore of scientific discussion. This new priority then aids in understanding the more recent advances that have been made. The unconscious is just such a priority. The concepts of psyche and spirit rest on it, as the determining factor with which they stand and fall.

The core of the new movement—occasioned by *the philosophy of the unconscious* in connection with a theory of organic development—is simply the old question concerning the nature of the soul. But, the complete scientific depiction of the unconscious must first undergo a rigorous process of purification before the concept of the psyche will find the complement it has been lacking; only then, at this higher phase of its development, will the concept of psyche determine decisively the form to be taken by the new philosophical system that has already begun to signal itself.

In order to speak of the unconscious, we first have to become conscious that an unconscious exists and that our consciousness manifests via processes that are operating unconsciously within us. In this way, consciousness, as the middle term, yields the unity of self-consciousness with the unconscious, in such a way that by spirit we understand the psyche having become conscious of itself, and by psyche we understand the spirit latent in the unconscious.

The human being tends quite comfortably to attribute that which makes him what he is—namely, his self-consciousness—to animals at the lower end of the scale and to the absolute at the higher end. But he falls into self-deception believing it is possible, both linguistically and materially, to step out from his own nature. On the other hand, in every other human being he is always encountering himself, for in others he rediscovers a disposition to consciousness that is identical with his own, which entitles him to make judgments and form conclusions about himself. Every other human being is his fellow human being, but no human being is the animal's fellow animal. It would be attempting the impossible to displace his own nature into the animal and to imagine its psychic life in terms of the qualities of his own consciousness.

Precisely because they are inaccessible to him, the human being lacks any appropriate expression for non-human conditions. If it helps him nevertheless to designate some external similarity by borrowing from himself—if only in order not to forfeit the advantage of the comparative study of nature entirely—still, he should not surrender to this deficiency in his language so unconditionally that image [*Bild*] and thing are utterly confounded. *The human psyche, the spirit, is self-definition.* On a side note, this is the basis on which all of so-called animal psychology needs to be corrected. The animal *feels* appetite and aversion; the animal *knows* by instinctive combination of sense impressions, which it then remembers. But sensations that presage representations, and representations from which concepts arise—these the animal does not have. How much more than this is human consciousness, which is conceivable only as the original disposition to self-consciousness. The human being makes up his representations of the nature of non-human "consciousness" at the expense of the integrity of his own self-consciousness.

Organ projection pursues the process of self-consciousness all the way to the world of culture, which has yet to be examined from this perspective. Organ projection is perfectly justified in making repeated reference to Carus's *Psyche,* given that *The Philosophy of the Unconscious* positively affirms its findings.

Let us not fail to acknowledge that, according to the degree to which it participates in the unconscious, it is above all through organ projection that we recognize the impulse human nature has *to reflect itself in itself.*

This human nature is the whole human being, the enfleshed soul whose first unconscious stirring is potentially already consciousness and spirit—though we must be careful not to assume that the concept of a higher and a lower, in the sense of an order of rank, applies to soul and spirit, unconscious and conscious, consciousness and self-consciousness.

Wundt refers to psyche and spirit as the same Subject: the former being the subject of experience which, conditioned by the boundedness of this experience, leads an external existence; the latter being the subject of inner experience, in which it is abstracted from its ties to a bodily nature. "This definition leaves entirely unanswered the question as to whether the spiritual is truly independent of the sensual. For one can disregard one or more aspects of a phenomenon without however disavowing that these aspects are present."[4]

The concepts of the unconscious and consciousness, as essential determinations of psyche and spirit, naturally also fall within the parameters of Wundt's careful definition—which, having resulted from rigorous psychophysical investigations, should not be underestimated. The degree to which the unconscious becomes the content of consciousness is the degree to which consciousness is self-consciousness.

If the infinite threads that interweave the human being into the entire universe are located in the unconscious, then all philosophy is actually also a philosophy of the unconscious[5]—and its ultimate outcome is our *self-conscious awareness of the unconscious*. Since the unconscious manifests in equal measure in the body and in spirit, self-consciousness is not merely consciousness of the subject of spiritual activity but also a consciousness of the bodily life that essentially constitutes the Self.

I conclude my own remarks on self-consciousness, so far as organ projection is concerned in this process, by quoting the following remark by Virchow, which we will do well to heed:

"The educated human being should know his own body, not only because such knowledge belongs to his education, but more so for the reason that, ultimately, the way the human being represents himself to himself becomes the foundation for all further thinking about human beings."[6]

There are singular remarks of enduring truth that alone suffice to attest to the genius of their author, even were he otherwise entirely unknown. The above remark belongs in this category. These admirable lines register science's noblest task—namely: an education that is unaffected by the barriers erected between expert and everyday knowledge; next, the ethical imperative that "the educated human being should know his body" as the means to this goal; and finally, the declaration of an education based on the knowledge of one's own body as the highest form of knowledge, for the reason that it is "the foundation for all further thinking about human beings" and, let us add, for *self-consciousness*.

CHAPTER 10

Machine Technology

The human being should know his own body. This means simply that he must learn to know himself. Our entire task is to describe the way in which he acquires this self-knowledge—that is, through the very concrete means consisting of the implements he creates with his own hand. Thus, what we need to demonstrate is that the integrated self-contained richness of a living union of limbs appears, outwardly dispersed, in the endless variety of discrete objects, as its counter-image.

The history of this discovery—of the immutable inner correlation between that union and these diverse items—is the history of the development of self-consciousness. We have in part brought this world of culture into lockstep with the gradually expanding perception of its connection to bodily being, a connection that rests on organ projection. And we came to a provisional stop at the point where, by means of the electrical telegraph, the play of thought reached the greatest degree of mechanically supported mobility it has as yet achieved. At the same time, we permitted ourselves a glance ahead into the field in which one may perceive not only the durability of fixed mechanical forms in space but also the duration in time of those configurations that escape confinement in the crudely sensuous because they are the means by which spiritual functions are manifest.

With respect to the generation, maintenance, and action of force, the steam engine was such a fitting image of the bodily organism that surface-level observation confused the distinction between compulsory mechanical movement and spontaneous organic movement, virtually confounding the representation of the machine and the human being.

Before we venture beyond the machinal transmission of thought effected by means of the telegraph into the field, referred to above, of more directly mediated spiritual communication, it is incumbent on us to seek absolute clarity regarding the concept of the machine. For in order to be able to answer the question whether the human being is a machine or not, first and foremost we have to know what a machine truly is. In short, the concept of the machine must be ascertained in order to be able to measure against it the representation that we make of ourselves.

We have only very recently been afforded the possibility of doing so, since Franz Reuleaux fully developed the concept of the machine in his work on *The Kinematics of Machinery*. The book introduces a genuine science of the machine and puts this in context with the other sciences.

The complete title of the work is: *The Kinematics of Machinery: Outlines of a Theory of Machines* by Franz Reuleaux, Professor and Director of the Berlin Königliche Gewerbeakademie. With an atlas and numerous woodcuts printed in the text. Braunschweig, 1875.[1]

The findings of this theory of machinery, which are in and of themselves of enormous value for the correct conception of the cultural idea, are particularly useful to the progress of our own investigation. The following reflections on machine technology—even in the form of a critical report, as they first provisionally appeared in the second volume of *Athenaeum*—are fitting here, as the connecting link between the two dominant spheres of the theory of organ projection that we have repeatedly identified.

Among the many phenomena that currently, in all fields of human activity, are urging us toward a new worldview, Reuleaux's work is of exceptional significance. For the new worldview is the new philosophy, which is called on to establish a proper balance between empirical and speculative elements.

For a long time, speculative systematics on the one hand and aphoristic empiricism on the other formed two opposing camps.[2] While speculation appeared to be expiring under the increasing transcendental hollowing out of that which nourished it, empiricism, overladen with realism, lost insight and ordering perspective on the mass of its accomplishments. The former watched as the shadows cast by verdant branches grew alarmingly thin, while the latter in its turn could hardly see the forest for the trees!

The empirical sciences' demand for restitution against the long-standing supremacy of the idealist principle was all too amply rewarded. Philosophy was summarily dismissed, and a decidedly materialist tendency proceeded to dominate—precisely by using spirit's own weapons against it, those of language and deduction. The materialist tendency managed to engage its opponent so intensively that, affected by the materialists' incontrovertible truths, philosophy's own transformation into *Naturphilosophie* went unremarked. For the moment, this is the preliminary peace that has been achieved between natural science and philosophy, since the latter freely admitted its need for continuous supplement from the empirical sciences.

One has therefore grown so accustomed to seeing political, religious, and social factors, the creations of art and the discoveries of natural science, as the driving forces of spiritual progress that precisely that empirical domain, which has always been the most indispensable ground for our entire cultural development, was almost completely overlooked.

If the sensory stimuli and impulses that come from the natural world outside and continue on inside the human being to become sensations and representations—if these indeed shape the inner world of consciousness and of spiritual life in general, then the question arises: why does an approximately similar development not take place in the animal kingdom? Animals too are susceptible to sensory impression, regard the same outside world with open eyes, only they persist in an endless stupor vis-à-vis the same.

We can answer this question, as was shown above, by specifying more exactly what in fact counts as the outside world. Typically, one might consider the outside world to be simply those things existing in nature, by which the human being is surrounded. But is there not another, more superior world on display beside this natural world—namely, the technical apparatuses proceeding from and epitomizing the human brain and hand? World Expositions offer an approximation of the quality of such things, while their quantity—distributed across the inhabited globe in the form of tools, weapons, buildings, garments, works of art and literature, equipment, apparatuses, and machines—is incalculable.

This latter is the outside world in which the human being has contrived the external continuation of himself, without which the understanding and appropriation of nature would be impossible and any explanation of his own being would be inconceivable. The human being becomes conscious of the artifactual outside world, just as that world, having advanced from the first crude tool modeled after the natural organ, is culminating today in a wealth of the most complex machinery. It appears to him as a world proceeding from himself, as an exterior that was once his interior. In this sense, the human being distinguishes the artifactual from the rest of the outside world, from telluric and cosmic nature—for the latter's presence [*Vorhandensein*] to the living being is merely the condition of its animal existence.

But nature, as raw matter brought into contact with the human artifactive drive, undergoes a refinement. Transformed into artifact, it comes to represent that other outside world. Belonging now to the domain of technical apparatuses beside and opposed to natural forms, it supports and maintains the cultural process—a process of growth through the reciprocal interplay of bringing forth and having been brought forth.

Among the works of human artifice, the machine towers above all. It is the subject of Reuleaux's book. Having shifted the focus of his investigations into *kinematics*, he founded a new discipline, which led him to title his book *The Kinematics of Machinery*.

This book expands on and develops in a systematic context a series of papers published in the course of recent years in the *Berliner Verhandlungen* under the rubric "Kinematic Communications." Even when first published serially, the papers were met with a warm reception in the majority of German polytechnical schools; as a complete work, they are bound to find their way into university lecture halls.

Given that the work is "not designed exclusively for study by experts, but also for those indirectly affiliated with the field" and intends to make the theory generally accessible, it breaks with the circle of specialists, appealing instead to the judgment of the mindful reader. In this way, the work promises to enrich the entire spiritual domain, from which this particular discipline has long seemed remote and remained estranged.

Earlier approaches to machinery consisted first of all in *machine theory*, which describes the nature of existing machines; *mechanical technology*, which depicts the particular fixtures by which the machine absorbs and deploys the action of natural forces; and *machine construction,* or *theory of design.* With the advent of kinematics, or the theory of mechanical gear trains, these formerly isolated angles on the machine have coalesced to form a complete view of the object.

In taking a genuinely deductive approach to the machine, the author intends, on the one hand, to bring about in the machinist a complete understanding of the mental operations that he is carrying out and, on the other, to replace an indeterminate, often incidental point of view with one that is both definite and scientific. Thus, what is indeed new in this book, according to its author, has to do "with the disciplines of logic and philosophy." Generally speaking, a certain disrepair in the state of philosophy may be at fault for its earlier, fragmentary approach to the topic of the machine, in consequence of which practical as well as theoretical machine experts tended to neglect philosophy or mock it outright, and scientific logic along with it. The exorbitant accumulation of figures and formulas needed reining in; toward this end, the aim was to revive a practice of the *independent treatment of the special case on the basis of general laws.* This alone enables, in place of the earlier theory of individual machines, the formulation of an actual theory of machinery—in other words, a machine science.

The hallmark of every discipline is its governing idea, in which is included the determination of the object; the object's determination endures solely because its concept is based on it. This *concept* of the object is the inexhaustible source of thoughtful reflection upon it. And therefore, the concept of the machine that Reuleaux derived seems to have yielded as if on its own the principles of its discipline. As machine science, the discipline is united in solidarity with all other sciences, according to a reciprocal give-and-take, its own persistence, and progress guaranteed by their prosperity.

We therefore find in this an "attempt to restore the link between machine science and the aggregate of all other sciences—a link that is on the brink of disappearance because of the drive toward specialization"—accomplished so strikingly that its benefit to the deepest interests of humankind can hardly be assessed.

Tools and machines neither grow on trees nor fall like manna from heaven; instead, they exist *"because we have made them ourselves."* As products of this Self, they bear the clear imprint of a spirit that is sometimes unconsciously finding, other times consciously intending. Therefore, in referring back to the site of their nativity, they explain and inform us of the organic activity to which they owe their emergence, as the after-image does its prototype. We must then regard them as crucial factors in epistemology generally speaking and in the development of self-consciousness in particular.

If the task of philosophy is to propose an answer to the question of the relation between human thought and the outside world, including the outer world of human artifacts, then a depiction of the emergence and perfection of the machine demands our most careful attention—especially if it is to function as rigorous scientific confirmation of Virchow's claim that "the same skill with which the human being achieves mastery over nature and, with it, mastery over himself plays an essential part in the development of conscience and in the moral upbringing of the human species."[3]

Were empirical objects understood as correlated with spirit's own-most purposes, as we have suggested, then skill and knowledge would authentically rise of their own accord to the order of a science.

And this is exactly what we see unfolding in a wonderful way in the present case—that of Reuleaux's book. An overview of its layout and execution should be sufficient demonstration.

The Preface and Introduction are concerned with the emergence of substance and the ideas that guide its ordering, as well as with the distinction between previous ways of interpreting and the ideal that should be aspired to.

The first several sections are well suited for general comprehension, since for the most part they deal with logic and philosophy, while the content of the remainder of the book is addressed more directly to specialized technicians. Given the existing antipathy to anything new or unfamiliar, specialists require a factual explanation before they can be convinced that the arguments they are being asked to follow are in fact being made by a thinker in full command of the details of their own specialties. Only then can a deductive approach to the object that is being undertaken for the first time take up the discussion with the weight of authority.

Reuleaux does not presume that the layperson is barred from the study of machines; on the contrary, he acknowledges that today's schooling provides most readers with enough amateur knowledge and experience handling tools and machines that educated persons should not have any difficulty in grasping the context and understanding the whole. Now we will take a closer look at the problem of the machine and follow Reuleaux's stepwise solution of it, which is possible only from a kinematic point of view. This is because the greatest value that even the most elaborate apparatus can have, if it does not move or operate, is the value of a model; only the act of movement legitimizes an apparatus as a machine.

The concept of the machine is built on a distinction between the external, sensible forces of nature, as those contributing to movement, and the latent forces of nature internal to the machine parts that produce resistance. The interaction of these sensible and latent forces, arranged so as to produce an intended movement, forms another aspect of the machine concept. Unlike general mechanics—which investigates motion as such, in response to the forces of nature—machine mechanics is concerned with movement that is strictly confined to a limited circle of means. In this way, machine mechanics delineates itself as a distinct discipline within the general field.

While earlier machine theory assumed change in location as given in the machine, now the question has become what *causes* this. *"As the science of the machine's fixtures, kinematics is concerned with systematically understanding the fixtures that determine the reciprocal movements in the machine, insofar as these movements are changes of place."* The basic feature of the machine is the assembly, not so much of individual elements as of bodies paired and belonging together, or *paired elements*. Of these, one provides the envelope-form of the other, so that when one is held still, the other remains movable, *but only in the unique manner that is specific to the pair*, as one finds, for instance, with nuts and screws. The movement of one element behaves *relative to* the affiliated element, but *absolutely* with respect to the fixed point in space (Figure 28).

From the reciprocal linkage of paired elements emerges the *kinematic chain*, whose members—that is, both linked elements, each from a different set of paired elements—are forced, as soon as a relative movement takes place in the chain, to carry out specified relative movements in a *closed chain* (Figure 29): "The chain consists of four like pairs *ab, cd, ef, gh*, each formed of a cylindrical pin fitted into a corresponding eye, and each

Fig. 28.

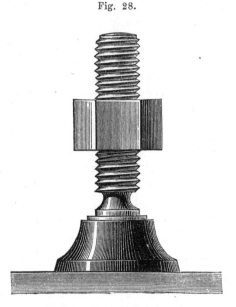

Figure 28. Machinal paired element.

Fig. 29.

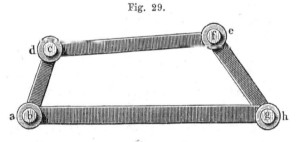

Figure 29. Kinematic chain.

positioned parallel to the others. Each link describes a circular movement relative to the neighboring link. Each rotation of *ha* relative to *gf* triggers a change in location of *bc* as well as *de*—therefore the chain is closed."

If, then, one link of the closed chain—which should have as many paired elements as it does links—is held stationary, then its relative movements become absolute. Such a chain is called a *mechanism* or *gear train*. In Figure 30, the link *ah* appears fastened onto an adequately fixed pedestal, such that, kinematically, the pedestal and *ah* form a single piece. "The movement that can now take place in the chain is indicated by dotted lines drawn between the main positions. This is the movement one recognizes as occuring between the 'beam' and the 'crank.'"

"A kinematic mechanism moves when a force is applied to one of its movable links such that the position of the link changes. The force thus applied carries out a mechanical operation that takes place under specific conditions. *The whole of this is the machine*" (Figure 31).

Fig. 30

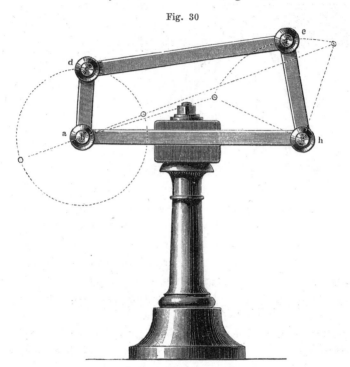

Figure 30. Mechanism, or drive.

The concept "machine," then, has to pass through a series of developmental stages. From kinematic contents like these—the paired elements, the closed chain, the mechanism—an inductive process gives rise to the machine, which is again resolved, through deduction, into its component parts. If the machine is more of an unconscious discovery, arrived at through largely obscure inductive processes, then "deduction and analysis are the means that lead to conscious induction and synthesis of the invention."

Fig. 31.

Figure 31. Machine.

By this, however, we in no way mean to assert that invention could somehow be determined in advance. Rather, for one who has acquired a fundamental understanding of the existing mechanisms, the ability to create the new proceeds from this understanding of the old. A fundamental understanding develops, step-by-step, from the elements of which the machine is formed to the realization of its complete concept—in other words, understanding is achieved, as always, by following the path of development that the object in question itself has taken.

At this point, we need to be sure our understanding of certain details is as clear as possible, for the sections to follow will build upon the theorem of paired elements and the kinematic chain. This is a bridge, initially accessible only to the expert, by which to introduce the reader to the *developmental history of the machine*. This should be strictly distinguished from the *history of the machine*, because it is the *development idea* itself that is the inexhaustible source from which the new discipline derives its evidentiary force and by means of which this chapter itself opens up a wealth of creative ideas and previously unimagined views, both forward- and backward-looking. It is the development idea that, relocating the emergence of the machine in the human spirit, leads to the domain of the spirit's historical development. The latter's beginnings date as far back as the most remote ethnological episodes, which means we have to rely on historians of language and on geologists to provide us with information about it. Reuleaux himself also credits the linguistic historian Lazarus Geiger with having identified the beginnings of the machine in the friction fire-starter, or friction drill, in keeping with the fact that culture begins when fire is made (Figure 32).

Fig. 32.

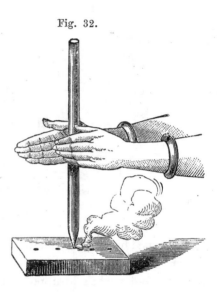

Figure 32. Friction fire-starter.

Therefore, the friction fire-starter's twin pieces of wood, in their whirling rotary motion, are the very first machine, or the first apparatus to have truly earned the name. It was subsequently discovered that a cord could be wound around the upright piece of wood several times and its ends pulled back and forth with the hands. In this way, through the mediation of a spindle, a whirling motion is conferred; thus the whorling drive, with its back-and-forth rotation, gradually enabled one to produce a continuous, non-intermittent movement (Figure 33).

The *undershot waterwheel* should be considered the first representative of continuous motion. Precisely these revolving motions were represented in *cart wheels* and the *potter's wheel*. Since the rotation occurring in the latter requires a machinal mounting for the piece to be rotated, the potter's wheel paved the way to the lathe.

Fig. 33.

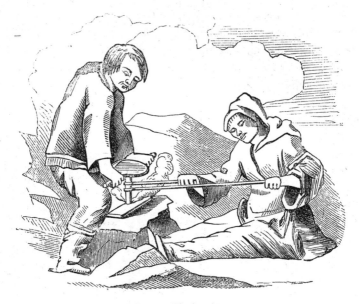

Figure 33. Whorling drive.

Thus we find here fittingly illustrated the nevertheless quite difficult transition from a whirling movement to continuous rotary motion by means of a cord or a belt drive and, moreover, from the narrow crossed belt and the cord looped several times around both rollers to the single, uncrossed loop. We are presented as well with a historical perspective on these transitions, from the apparatuses of antiquity to today's continuous belts and wire cable drives.

The use of a cord drive to move a spindle—whose original rotation by hand has not yet disappeared altogether—generated the *hand spinning wheel. Ropemaking,* in the broader sense, also belongs to spinning. The loom that works spun yarn is, like the bobbin, not a machine in the sense indicated, but an apparatus in which the actual machinal motion is only rudimentarily present. Here the investigation segues into the up-and-down motion of the seesaw, or the Hindu *kuppilai* that was used to scoop water; Reuleaux also calls attention to a similar apparatus that the Chinese used with a rope drill in order to dig artesian wells. The *bow* and *arrow* also take part in this rectilinear motion. The velocity of the *bow,* the *crossbow*, the *ballista*, and the *catapult* derives from an elastic force-collector. The blowgun, according to its mode of projectile operation, is an antecedent stage of the powder-powered gun.

Reuleaux's comments on the paired element of the primitive screw and the primitive nut are less decisive with respect to the emergence of the screw generally and with respect to the priority of the right-handed screw and the manufacture of the nut or the wood screw. The obscurity here may be clarified with the help of physiology. Namely, because the left side of the body—by nature, the side where the heart is located—is more sensitive to stress or too strenuous leftward movements than the right side is, this may explain the right hand's priority; and therefore the screw thread, obviously the continuation of the rotational movement away from the body, had to be a right-handed one. As far as the nut is concerned, indeed every hollow thread thus provisionally produced by the rotation of a screw in wood is the primitive nut, however imperfect—an assumption that is by no means inconsistent with the assertion that the drill whorl obliquely indicated the way to the screw pair.

With respect to the relation between *motion* and *force,* it has been suggested that children and primitive human beings are more impressed at the sight of the movement of windmill arms, waterwheels, and stampmills, for instance, than they are by the force these can generate. We can take this as evidence that, slowly and gradually, from the increasing varieties of machinal movements emerged the generation of immediately accessible force and not the other way around. For this reason, it is not the lever but the fire-drill that can legitimately claim to be the first machine.

While the human was more or less unconsciously applying the power of his limbs when he moved the first machines, the difficult intellectual operation of distinguishing conceptually between force and motion took place by degrees. Ultimately, the human being went on to develop the motor aspect of the machine in such a way that he was able to substitute natural forces, animals, water current, and wind for his own muscle power in the operation of early machines. The principle of collecting the muscles' sensible forces—which we have already alluded to in the collection of force in the elastic bow—"later extends, without discrimination, even to elementary forces, and today we find it applied perfectly, from the tiny spring-works of the pocketwatch and the gunlock, through countless tension mechanisms, to Armstrong's hydraulic accumulator and the air vessels of the Mont Cenis borers.

"The discovery of the motive force of steam occurred late, long after certain quickly flammable or explosive substances had been discovered—though in these we find only the latent force that nature has amassed in enormous quantity in its decomposable matter. In such substances, the human being discovered a source of power the magnitude of which he did not at first suspect; when set to work in the machine, however, these substances have helped to lift him far above nature and set in motion the greatest revolution ever witnessed in the life of humankind."

Once the relation between force and motion is clarified, we reach the nodal point of the investigation, where the question of the properly kinematic feature of the perfected machine is decided. This has to do first of all with the form in which a paired body completes its kinematic connection.

In the case of most waterwheels, any vertical motion of the axle from the uncovered bearing is prevented by the weight of the wheel itself. The wheel's weight is therefore the force that holds any expected interfering force in equilibrium, insofar as it closes, so to speak, the position of the envelope-form that is otherwise left unclosed. Therefore the designation of *force closure*.

On the other hand, for paired elements with complete envelope-forms, as is the case with nuts and bolts, for instance, any interfering action by sensible forces on the form of movement unique to the pair is countered by forces latent in the pair. This closure in complete paired congruence Reuleaux calls *pair closure*.

Progress toward the machine's perfection now consists *"in a decrease in the use of force closure and its substitution by increased pair closure and the closure of the kinematic chain formed of these pairs."*

Once the process of replacement of force closure by pair and chain closure is ascertained through a look back to the machine's beginnings that we have just described, Reuleaux moves on to a description of *modern machinery*, which he dates back to the invention of the steam engine. Here he shows that this process of substitution should be regarded as the essential form of the machine's development, albeit now greatly accelerated.

The cart, the turbine, agricultural machinery, gear wheels, shipyard cranes, mill works, and especially the steam engine show that "the force closure limitation was essentially the means by which machines have been made adequate to better carry out the portion of the task assigned to them." As a result of the effort to translate the makeshift results of the first experiments into self-acting paired elements and simple drive mechanisms, the machine began to take on a greater portion of any given task and occasioned the invention of new mechanisms. The great variety of force closure mechanisms, as a transitional stage characterized by machines invented by pure empiricists, began to diminish in part under the hands of the skilled engineer. Continuous improvement through substitution wherever possible of the uncertain with the certain—that is, of force closure with pair closure—is the basic feature of the earlier machine period. By contrast, "the sudden, striking emergence of already quite perfect machines" is what characterizes modern machinery.

The reflections that Reuleaux at this point ties in with his depiction of the inner essence of the machine permit a deeper look into the brilliant conception of his science—its heretofore unsuspected inner correlation with history's greatest tasks opening up for us a world of surprising new realization. And, moreover, the book is rousing, which betrays the fact that Reuleaux's treatment of an otherwise dry and impassive object came to be under the aegis of a deeper insight into art and general science.

The book's greatness lies therein that, put briefly, it has made a case for a metaphysics of the machine. Through all the rattling and clattering, creaking and thumping of wheels, rollers, and hammers; through the hum and whirr of the loom; through the locomotive's whistle and the ticking of the telegraph, we are reminded that here, in these mechanisms, each of which recalls its organic prototype, the essence of the human being itself has become objective. We admire Reuleaux's grasp of the governing role of the unconscious, even in the genesis of the industrial world. To those who mistake this, his rebuke is implicit, when he writes:

"We must not overlook that the general development of the machine has previously been accomplished *unconsciously,* to a certain degree, and that this unconsciousness in the old mode of production has left its particular mark—albeit one that eluded our perception. But the mode of production of modern machines stands in contrast to this, given over from the outset to practiced hands, as mentioned above. Already we have captured some if not many things clearly and deliberately! By now, we no longer see improvements to apparatuses on the old basis of some deficiency; rather, we find brand-new devices brought suddenly to life."

Here, as elsewhere, Reuleaux testifies to a curious trace left by the evolution of the machine, a process taking place differently depending on whether unconscious impulse or conscious deliberation exerts a greater influence. With respect to the older, graduated mode of production: "For this reason the first machines result from the inexperienced hand of humans, whose power plays a subordinate role, as they do not surpass the more or less *unconsciously* practiced efforts of the limbs." Furthermore: "The theory of machinery furnishes evidence of a coherence among the results of diverse thought processes. These cohere in a single field in which one proceeds *unconsciously*, therefore slowly and by detour, according to laws that one cannot escape, precisely because they are true laws." In discussing the analysis of machines, he remarks: "Our analysis has led us to the governing idea that also guided all those who have *unconsciously*, unwittingly invented or improved machines." The unconscious is always in evidence wherever the invention is ascribed to a kind of revelation, attributed to a higher "inspiration" that is visited upon the genius in his workshop of ideas, when it is ascribed to "an obscure induction" or to "feeling and instinct." These passages about the unconscious may help to clarify the expression we have chosen, the "metaphysics of the machine." Even though they do not directly reference the *philosophy of the unconscious*, because these passages have arisen directly from investigations in a particular discipline, they have all the more value for such a philosophy—the inestimable value of unbiased agreement and objective confirmation.

Accordingly, if we keep in mind the principle of force and pair closure in the development of the machine, and if we locate, generally speaking, force closure on the side of unconscious seeking and pair closure on the side of conscious invention, then, following Reuleaux's explicit claim, the antagonism between force and pair closure in the machine will never quite be eliminated. The concept of the mechanism rests on this: it cannot do without control by the power of the human hand and mind if the machine is to be the image of the living, organic principle in an otherwise lifeless mechanical formula. And so, psychology and physiology are faced with the task of establishing that the principle that creates an entire world of culture and the principle that is immanent in human nature are indeed one and the same. This task needs to be accomplished on the basis of the insight that this antagonism in the outside world is dissolved in the monistically mediated distinction of phenomenal forms in our inner world. In the future, then, force and pair closure shall find their psychophysical value, as the machinal idea existing in the world of culture is called on to shed new light on the nature of the human psyche and on the world of spirit generally—which it does with explanatory reference back to its own primal ground.

The sublime observatory is thus erected, offering us a comprehensive view. With regard to the cosmic forces acting independently of one another, and with regard to the molecular forces concealed in the existence of the machine that counteract the action of the motor, and remembering as well that force closure is the form in which a remnant of cosmic freedom is admixed in machinal systems, a transitional zone is indicated—which leads from the ideal-machinal system into the cosmic. Let us add that inhering in the cosmos is the microcosm and that the human being himself, in life and limb, represents the ideal-machinal system!

A powerful impulse to this realization is motivated by our reflection on those phenomena that parallel the inner nature of the machine—particularly in the arena of human behavior, which the development of machinery is the intensified counter-image of, in that the machine, via the artful restriction of movement, compels the action of forces toward a single, intended goal.[4]

We have already touched generally on the relation of motion and force. At this point we want to discuss once again the "impetus to machine development" as the need to generate movement and as the need of force. Irrespective of which of the two—force or motion—originally took precedence in the machine's development, no matter whether they were more or less unified or distinct, whether they cooperated or operated in parallel, whether one sort of machine operated with an abundance of force and another with a great range of motion, what matters is: "Warfare and architecture and cargo transport in general drove the expansion of deployable force; manufactory and time-keeping instruments, among others, required an increase in the scope of realizable movement."

Before the human being can individualize cosmic forces for his own purposes, first he must distinguish or *discover* these from among the totality of collateral phenomena, through observation and natural scientific research. Like the generation of force and motion, discovery and invention go hand in hand, and *"with the discovery of each new source of power comes the invention of means for exploiting it."*

The core idea and innermost guideline for the progress of the machine lies in the replacement of force closure by kinematic closure. Reuleaux recognizes this as the task of polytechnic education. Were a course of study to orient itself along these lines, we would succeed in part in retaining, in part in restoring in the builder of machines a common affection with the totality of all practical mechanics and, moreover, with the totality of human activity in general.[5] We would also succeed, following the popular battle cry for the "division of labor," in limiting the excessive division of knowledge to boundaries within which it would remain possible to reaggregate the divided fields and return them to a higher unity, under the direction of the development idea, wherein lies the strength of contemporary science.

The course of the machine's development that we have followed thus far—from the lower and higher paired elements, all the way to the kinematic chain—yields such a wealth of forms and a diversity of derivative and inferable instances that it becomes increasingly difficult to maintain both the clarity of and specific language for individual characteristics. This problem, however, is remedied by Reuleaux's "kinematic notation." Once its concepts have been scientifically determined, its application should follow a similar course as that of mathematical and, after it, chemical notation.

Literal and semiotic abbreviations like these are all the more expedient, the more their stenographic character conforms with the conventional international scientific terminology. In studying this notation, which Reuleaux has already put into practice in his book, the reader has the sense that here a great kinematic feat has been accomplished. Reuleaux's own both clever and instructive notational system operates as an analogon to what he expects from the machine builder, as an inventor: "In this way, one is spared the need to return time and again to previously defined properties. Given the economy of expression, one is able to reach conclusions concerning the coherence and reciprocity in the interconnected totality that would hardly be obtainable and, moreover, barely communicable using the typical mode of expression."

In this way, the investigation itself produced its own tool with which to cut a path through the thicket of "kinematic analysis" that follows. At the same time, this notational invention substantiates its own utility in the further course of the investigation that initially gave rise to it. Earlier force closure-like wildlings of notation—among others, Charles Babbage's *On a Method of Expressing by Signs the Action of Machinery*—are best replaced, if the comparison is permitted, by this pair closure-like system.

As far as *kinematic analysis* is concerned, its task consists first in decomposing a kinematic apparatus into those parts that are to be considered kinematically as elements and then in determining the arrangement in which these parts assemble to form paired elements and kinematic chains. While within this broad sphere Reuleaux undertakes a series of investigations, he begins with the so-called *simple machines* or "mechanical powers": the lever, inclined plane, wedge, pulley, wheel and axle, and screw, among which only the lever, the inclined plane, and the screw count as paired elements, or closed pairs. He then delves with astounding thoroughness into the vast field of crank mechanisms, chamber-crank chains, chamber-wheel gears, and the constructive elements of the machine. It is left to the experts and specialists in the school of applied theory to determine whether and to what extent kinematics may be said to have succeeded in transforming "invention" in the familiar sense and in replacing it with a scientific method of development. Our discussion, however, according to our intention laid out at the start, is limited to marking out those points around which new general scientific relations may be grouped, and we turn now to the chapter on "The Analysis of the Constructive Elements of Machinery."

This section is the most accessible to a general readership, and even the non-expert will not be perplexed, for he will find a medley of many things here, the forms and names of which will be familiar to him from ordinary life. Now, though, he will find these things newly reordered, clearly and conceptually, into three series, those being *rigid elements, flexional elements,* and *trains.*

The rigid elements are either *joints* (for forming links): rivets and riveted joints, keys and keyed joints, strained joints, screws and screwed joints; or *elements in pairs or in links*: screw and nut, pins, bearing blocks, shafts and axles, fixed couplings, simple levers, cranks, compound levers, connecting rods, crossheads and guides, friction-wheels, toothed-wheels, fly-wheels.

The flexional elements are separated into *tension organs* and their chain closure application: belts, belt drives, cords, cord drives, chains, chain drives; into *partners of pressure organs*: pipes, pumpbarrels, pistons, stuffing boxes, valves; and into *springs*: tension springs, pressure springs, bending springs, twisting springs, strut springs.

The trains consist of click gears (the simplest), brakes, detachable and movable couplings.

The three modes of closure, which make up the transition to analysis of the complete machine, are *the normal constrained closure*, where all of the relative movements of the links are determined; *the incomplete or unconstrained closure*, where the relative movements between links are indeterminate; and *the fixed closure*, which cancels the reciprocal movement of the links.

The purpose of this detailed survey of the constructive elements of the machine is to provide, as an example at least, a general picture of the genetic method in evidence here, as it is throughout the book, even if in a dry sequence of terms. For surveys of this sort, should they otherwise be accorded the merit of an internal consistency, must, as parts, correspond to the development idea that permeates the whole, as a leaf corresponds to the plant. And so, with the reordering above, the difficult task of "detecting what is resolutely lawful in the complicated forms built of individual parts, and separating this from what is accidental" should be regarded as accomplished.

The architect who wants to erect a building has both in his head and on the piece of paper that holds the draft of his plan everything the construction requires—foundation and floor, the various materials and their formation, human labor power, animals and technical apparatuses, reference to weather and time of year, to sun, air, light, and a hundred other things and considerations. And the architect always proceeds from the whole before he deals with testing for requirements, with initiating and executing the detail. The result of the physical work, realized in the configuration of raw material, is the same entire house that was already there in the beginning, in his spirit, only now of course it is sensuously present.

The same is true for scientific procedures: what comes at the end of the investigation is always what was there at the start, albeit augmented by the addition of a lengthy spiritual operation. On the one hand, we have the house as it stands on paper and the house as it rests on the built foundation; on the other, we have the material of thought, existing as an open question, and the same material of thought, scientifically built out at the end of a process of conceptual development!

Just before his *analysis of the complete machine*, Reuleaux himself completes "the circle which was begun in the first chapter, where, proceeding from the complete machine, we undertook its decomposition in order to examine it from those points of view we have identified as belonging to the kinematic sphere." First he tests the tenability of the conventional interpretation, then divides the content of the complete machine into three groups: *receptor, communicator,* and *tool.* "The receptor is that part on which the natural motive force acts directly and to which is conveyed the mechanical work as it becomes available; the tool delivers the work in an appropriate manner to the body to be processed; between receptor and tool, whose movements in the vast majority of cases do not coincide, is the communicator, the mediator of motion." Examination then shows that all three categories *could* be present and clearly recognizable in a single machine, though they cannot logically be described as essential parts but only "accidental members." For then it is shown that in certain machines a tool is either only very faintly recognizable or not present at all; furthermore, that receptor and motor—whether water, wind, steam, gas of another sort, weights, springs, an animated being—always enter into a kinematic pairing or chain in which the actual receptor is difficult to distinguish; and that, finally, in other cases the concepts of communicator and tool are confounded to the point of complete indistinguishability. Therefore one cannot discern a tool in the steam engine. Therefore again: if, after the clock weight has been removed, the cord or the chain is lengthened to equal the heaviness of the missing weight and the clock is meant to operate by the cord alone, then the weight cannot be the motor, nor can the cord be the receptor, since its quality has not changed. It is likewise unclear, in the case of the thread in the spinning machine, where the communicator ends and the tool begins.

This then leads to the consideration that machines without recognizable tools—cranes, locomotives, steamboats, clocks—are tasked with effecting a *change in location*, with transporting; and, furthermore, that machines in which a tool is actually present—lathes, planing machines,

bolt cutters, bandsaws, and so forth—characteristically subject the body
being processed to a *change in form.* Leaving aside the fact that in several
of these machines a change in form is associated with a change in location—as, for instance, the mill transports bodies even as it decomposes
them into parts—Reuleaux separates machines into two major classes,
vis-à-vis their purposes: *machines for changing location,* place-changing or
transporting machines; and *machines for altering form,* form-changing or
transforming machines.

With this, Reuleaux delivers his verdict on the conventional threefold
division of the contents of the complete machine into receptor, communicator, and tool. And if we add to this "that the concept of the tool
is not actually the machine's originating concept, rather it is an incidental feature of the machine and therefore cannot serve as the basis for
an understanding of the complete machine," then the concept of "tool"
is not depreciated but rather liberated from a distortion that is damaging
to the clarity of the machine concept.

For, with reference to its original character as *hand tool,* the "tool" not
only represents the very possibility of machine formation, but it also
includes—in its broadest definition as both *means* and tool, as the mediator of cultural ends, as the begetter of work in general—all machines,
whether or not they can be said to contain particular tools.[6] In this sense,
the locomotive, in which there are no particular tools, is itself the tool,
understood as an entire machine. Conceiving thus freely of the tool concept as a universal cultural power, the concept is secured against false
restriction. Certainly, it was the effort to appropriately delimit his otherwise comprehensive study that led Reuleaux to exclude entirely the hand
tool from the beginnings of machine development. A limit must be drawn
somewhere. Yet: because the hand tool refers back to its origin in the
hand itself—which the ancients deemed the "tool of tools," or ὄργανον
τῶν ὀργάνων, an expression that aptly convolutes the concepts of tool
and organ—further interpretation of this relation allows us to see the
affinity between a living organic prototype and its lifeless mechanical
after-image, an organ and its technical extension.

When analyzing the complete machine, the human hand is of course
considered in a very different light from this—namely, it is assimilated to
the machine drive as an organ. At the spinning wheel or the whetting
stone, for instance, the worker and his limbs are enlisted in the drive

mechanism, guided by its will power, such that the worker sets his own organic laboring body, as its own self-contained kinematic link, into a closed kinematic chain with the lifeless laboring machine. The machines powered by human hands and animals are hereafter "to be regarded as complete machines and need not be distinguished from those powered by elemental forces." For in all cases the motor enters into the kinematic chain as a link, in contrast with the older idea that placed it outside the machine.

Especially important are machines like the steam engine, the water-wheel, and the turbine, which serve as motors for the operation of a variety of other machines. These, as a category of machines complete unto themselves, are known as *powering machines*. Likewise complete are all place- and form-changing machines, called *working machines,* that are configured so as to be operated by a powering machine.[7]

As far as the human being as working machine is concerned, allow Reuleaux's "examples of the descriptive analysis of complete machines" to show that human cooperation in the machine diminishes to the degree that the machine's independence increases, and that the most perfect or complete machine will ultimately be the one that requires the human being only to initiate and terminate the machinal process. "In general, machines are noticably striving toward the pinnacle of perfection, indeed some are already within sight of it."

The machine's approach toward this goal has an enormous impact on society, given that the machine's development is connected with the totality of culture. The conclusion of the analysis of machines is concerned with this topic.

Large-scale industry's exploitation of the nearly unlimited power of steam has made possible mass work in colossal factory buildings; these in turn require masses of workers to cope with the amount of work to be done. In the struggle between the upside and the downside associated with factory work—on the upside, the inexpensive products of utility industries and the unlocking and utilization of great quantities of natural resources; on the downside, the devaluation of family life, the increasing price of life necessities, and the obvious decline in manual skill—the *"worker question"* arises.

As the magnitudes of power available to the steam engine increase, it becomes increasingly easy to manufacture by means of it hydraulic and other kinds of powering machines as well as all kinds of working machines. "Therefore this one powering machine, *the steam engine, as the parent of a legion of working machines,* is also the master of the situation." From this it follows that the remedy for the conflict between capital and labor is to be found precisely where it originated. If the large-scale machine stands under the warranty of capital, then the small working machines can, on the other hand, be produced at accordingly lower costs—thus opening up the prospect that the activities of the small-scale producer will be profitably revitalized and that cottage industry will find new sustenance. "Already small-scale industry is showcasing small working machines, above all gas-powered machines, hot-air machines, water-column engines, and, at a very promising experimental stage, petroleum-gas machines."

Reuleaux calls small motors *the true powering machines of the people* and affirms that air- and gas-powered machines, since they now operate considerably more cheaply, may confront the steam engine with some lively competition. He counts these among the most important of all new machines and locates in them *"the germ of a complete reconfiguration of a segment of industry."* Quite promising practical achievements are already in evidence.

The worker movement is a factor in this process of reconfiguration—a driving force and a means to progress that, once its service is rendered, will again disperse. We maintain, therefore, "that the manifest hostility toward human well-being is not contained in *the principle of the machine."* The centralization of labor, and the extreme tensions accompanying it, were provisionally necessary to be able to produce inexpensive machines in enormous quantity for the small-scale producer and will gradually return on its own to healthy dimensions. For true decentralization does not mean destruction of the center but production of balanced relations between center and periphery, between the cyclopic large-scale machine and the scattered small working machines. In the future, both will be mutually guaranteed, in the sense that the former is as indispensable for the vastly increasing demand for working machines as the manufacture of the latter is for the operation of the former. The threat of absorption of cottage industry and domesticity by large-scale industry is real. However, the revitalization of small-scale industry by machines will bring about an equalization and, with it, great progress toward a higher stage in civil human conduct.

The steam engine's transportation function in the mining industry as well as on rail- and waterways will not be subject to any significant change, since engine power does not tolerate division in such instances where it is effective only in its full extent and capacity. The steam engine's enduring sway is vouchsafed where it participates in the manufacture of other large machines and when at the same time sufficient limits are placed on the sacrifice of health and well-being. On the other hand, a much needed end will be put to the unfortunate circumstances the expanded fiber and textile industries are suffering, once the overly rigid concentration of industry inevitably relaxes.

The excessive concentration of vital energies at a single point of operation, to the detriment of all others, ultimately results in global degeneration. The therapeutic balance must reestablish itself as the surplus of labor power peacefully resolves, returned in part to a disadvantaged agricultural sector and in part to cottage industries, which will have been revitalized by new working machines.

Steam power precipitated the social tempest, and it alone can and will vindicate it.

It will be as if scales had fallen from the eyes of anyone competent to understand without prejudice the question of our time, when one reads: *"Wise restriction created the State; it alone preserves and enables its greatest accomplishments. Restriction in the machine enabled us to subdue the most powerful forces and to submit them to our management."*

If the analysis of machines taught us the properties of constrained motion which those connections formed of paired elements, kinematic chains, and mechanisms possess, then the *kinematic synthesis*, which forms the cornerstone of the theory, "allows us to specify those paired elements, chains, and mechanisms which, when properly connected, realize a particular kind of constrained motion." This task is considered most significant because its aim is the creation of new machines and therefore the continuing development of machines. But synthesis in applied kinematics should be one of the means of investigation, not the canon for how tasks ought to be executed.

The survey of the *synthetic* field that follows, illustrated with many examples, leads to the important discovery that the key accomplishments of machinery coalesce around a rather small number of kinematic chains and that, therefore, the field of kinematic problems is not incalculable.

At the conclusion of this brilliant work dealing with knowledge of the laws that govern the formation of machines, the express conviction is that "much of what there is to accomplish can be accomplished with very few means, and that the laws according to which we are to accomplish these things are readily apparent"—indicating that relation of primary affiliation, according to which each artifact, each manufact is an organic factor!

Now, when Reuleaux calls the organic being operating the machine a perfect machine, he does not mean it literally, but is using this language as an expedient to explanation. Language will often seize on a comparison either when, for the sake of brevity, it is better to avoid a more elaborate definition or when—since, as Jean Paul put it, the machinelike is nearer and more vivid for the human being than his own interior—one absolutely cannot avoid couching organic relations in mechanical terms. This passing shadow that fell on the organic being vanishes the instant that Reuleaux stipulates that the theory of machinery "must not disregard the machinal intervention of the human hand in finalizing what the machine produces—in other words, the *study of the living being* in its capacity as working machine." By this he is not calling on the theory of machinery to expand its field, though it does seem that the very idea of expansion in this direction could promise curious psychological discoveries.

Still, anything owing its emergence to the organic activity of human beings inevitably bears the stamp of spirit, appearing as organic activity and not otherwise.

According to Reuleaux, the primitive machines were all "devised of human wit—sometimes of such exceptional wit that they were prized as gifts from the gods. But they were always *devised through a process of thought that passed through certain stages.*"

Indeed, they were devised from the organ function immanent in thinking, whose fluent, animated norms and relations were unconsciously fixed in the form, number, and law of the lifeless material after-image.

That the unconscious originally accompanied machine formation—which Reuleaux notes on several occasions—explains the artifact's agreement, in part with the form of particular organs, in part with kinematic processes, depending on whether the organic movement is self-generated from within or the mechanical motion is imposed from without.

Given the operative restriction in the field of artifacts, we associate the term *kinematics* (from κίνημα, the moved) with the mechanism, whereas the more familiar *kinesis* (from κίνησις, the moving) is associated with the organism. Splitting the concept of motion in this way, into passive and active, may enable a satisfactory explanation of the relation between the organism and the machine.

First, however, we have to settle the question as to what extent organ projection complies with the concept of the machine that emerges from Reuleaux's work and, above all, whether the tool concept as we understand it agrees with the machine concept as given in *The Kinematics of Machinery*. As soon as we grasp that the machine is as much a continuation of the hand tool and of tools generally as the tool is the continuation of the hand and the organ, then we can answer the question affirmatively.

It does not matter whether it is the spindle proceeding from the rotating hand, and from the spindle the spinning wheel, and from this the spinning machine; nor whether it is the primitive millstone proceeding from human molars, and from the millstone the wind- and water- and later the steam-mill; whether, in progressing toward the complete machine, the immediate and continuous action of the hand diminishes— it does not matter because, throughout all of these transformations, the primary correlation remains unchanged. And, as it's said that an honest man won't "abon his might for a' that, an a' that," the human hand will never let go its grasp on the machine, no matter the degree of perfection the machine might attain.[8] The machine should not be thought separately from its origin; it would cease to be a machine if it were thus disconnected. The kinematic train is the actual continuation of the vital organic kinesis that Reuleaux sharply distinguishes, as the living working machine, from that which is lifeless.

The living working machine is at any rate a machine, to continue speaking figurally for a moment. But: it is the machine that was present before any machine formed by the human hand; it is the universal machine, the common archetypal and paradigmatic image of all the particular forms of machine technology, the machine *incarnate,* a kinetically jointed chain composed of paired limbs—in short, the bodily organism or *ideal machine,* its will both its inborn motor and the universal motor driving the totality of machinal creation!

Where each member of an indivisible whole is a rudimentary natural tool, then the whole appears as organization. The living *kinetic organization* of the organism is, realized machinally, a *kinematic chain* of pieces and parts.

While the geometric manner of depicting motion—or, phoronomics—is apparent in machine parts, organic organization precludes geometry entirely. Just as the one vocal organ in the human being, the pipe organ's prototypal image, is the condition of possibility for each of the pipe organ's individual tones; and as each individual pipe, depending on how its scalar expression has been arranged, can emit only *one* particular tone from among the multitude of tones that are unified in the human vocal organ; so too the one kinetic organization, the *one* organism, is the condition of possibility for each of the individual kinematic chains found in the machine. Each individual machine is thus distinguished from every other, according to both its material composition and its particular geometrical shape.

What the arm and the linked chain of ossicles in the ear are, in detail, in microcosm—that is the organization of the organism *in toto*. The hand then executes and translates to the sensuous reality of number and measure what is already present in this organization as possibility. The multiplicity of machine formation is in this way a true apotheosis of the universality of the *one* manual dexterity. The marvelous "army of machines" on display at the World Expo all originate in the hand, refer back to manufactory,[9] and from there to the spirit that leads the finder and likewise the inventor as well as the worker by the obedient hand, according with its—that is, the organic—prototypal image.

It is the hand that endows lifeless configurations with a glimmer of something human, that gives the impression there is something like a soul in the machine—an impression that once inspired LaMettrie to write *Man a Machine*. Having sat collecting dust for over a century, the book has justly found new advocates in figures like Albert Lange and, above all, du Bois-Reymond.

Without going into detail concerning the great truth as well as the great fallacy of this largely vilified book, we invite the reader to judge for himself, under advisement of a peculiar observation of Reuleaux's on the nature of the machine:

"Just as the ancient philosopher compared the ceaseless transformation of things to a perpetual flow and condensed this observation into the phrase 'all is flux,' we can sum up in the expression 'all is revolving' the countless phenomena of motion in that wondrous product of the human intellect, which we call a machine.

"For the practical machinist, who has familiarized himself with modern phoronomy, and more still for the theoretical one, the machine is animated in a particular way by all the rolling geometrical forms inside of it. Some of these make themselves physically evident, like the belt-pullies and friction-wheels of the railway carriage, for instance; others, like gear-wheels, remain slightly veiled; still others are drawn tightly together in the interior of massive bodies that give almost no indication of them; and finally, there are those, like the mechanisms formed of cranks and rods, whose patterns extend far beyond and encompass the bodies to which they belong, their branches stretching out to infinity, in outwardly unrecognizable shapes. In the midst of the noise and tumult produced by their bodily proxies, these forms silently accomplish their vital function of rolling. They are, as it were, the soul of the machine, commanding its bodily expressions of motion and reflecting them in a pure light. They are the geometrical abstraction of the machine and impart to it an inner significance, alongside its outer one, which gives it an intellectual interest to us much greater than it would have without them."

In this way, organ projection has a powerful ally in the machine. For the history of the machine's development—as a singular emanation of the general development idea, whose knowledge-producing power today more than ever stabilizes and orients philosophical research—is intimately related with the body and soul of the human being. Machinal kinematics is organic kinesis unconsciously transmitted to the mechanical; and learning to understand the original by means of its transmission becomes the conscious task of epistomology!

CHAPTER II

The Fundamental
Morphological Law

We are now fully informed about the phenomena of machinal kinematics. To understand the concept of organic kinesis is more difficult. Kinematics has to do with motion forced on a mechanism from without, while kinesis is spontaneous motion, organic self-movement. The nature of spontaneity is the great riddle confronting humankind. We therefore place an even greater value on those scientific achievements that permit us to distinguish between what we are ignorant of at the moment *(ignoramus)* and what we are certain to remain ignorant of forever *(ignorabimus)*.[1]

We count Adolf Zeising's theory of the proportions of the human body among achievements of this kind. His discovery requires our closer examination, since it illuminates for us one of the basic laws governing organic kinesis.

Last spring brought us the news of Dr. Zeising's death in Munich, on the 27th of April. Among artists and scholars, the deceased was best known for his work on *the golden ratio*.

The broader public, however, remained largely unaware of this book, since it was published at a time when the natural sciences were otherwise occupied with the most prodigious researches and discoveries that seemed to intervene directly in life itself. The public may then have been less compelled to turn its attention away from these immediate goals toward endeavors that at first glance seemed to have more of an idealist tendency. At that time, if natural scientific research did not signal an effort as paramount, then that effort was condemned to remain in the background for the interim, biding its time.

This interim is now over. Other than a few sporadic plaudits from the natural sciences shortly after Zeising's initial publication, what proved crucial for his theory was that two key figures—Gustav Fechner, the founder of physiological psychology, and Wilhelm Wundt, its preeminent exponent—spoke favorably of it, albeit within certain limits. At the same time, renewed acclaim for Zeising's theory came from the opposite direction, from speculation, which was then endeavoring to reconcile itself with empiricism. It is in effect philosophy, then, that principally sustains the golden ratio's law of proportion as regulating all aspects of organic vitality.

Under the auspices of "philosophical evidence," Zeising's law figures as the high point in which the latest work by Otto Caspari culminates. As one had hitherto vaguely suspected more than comprehended, Zeising's law is essentially correlated with Caspari's solution to *The Fundamental Problems of Cognitive Activity* and should also be put in close concert with social anthropology, for the sake of its own orientation.[2]

Contemplation of exemplary works of ancient sculpture had awakened the idea early on that master artists must have known the human body's law of proportion and obeyed a doctrine of formal beauty based on it. For this reason, there have been many attempts to bring this presumed law to light.

Adolf Zeising managed to do just that. We may regard his work as having solved the problem, especially if we take into account the findings of later works based on the foundations that he laid.

Zeising's major work—no one has yet refuted it—is titled *New Theory of the Proportions of the Human Body, on the Basis of a Heretofore Unrecognized Morphological Law Permeating the Whole of Nature and Art* and was published over two decades ago. Shortly thereafter, he published a work on *The Standard Ratio of Chemical and Morphological Proportions*.[3] Zeising discovered the fundamental law through application of an ancient Euclidean theorem concerning the dimensions of the human body (Figure 34).

This geometrical theorem, known as the golden ratio, deals with the partitioning of a line in such a way that the ratio of the smaller segment to the larger is the same as the ratio of the larger segment to the entire line. Zeising cannot say exactly where the term "golden ratio" originates, yet he thinks it likely has something to do with the extraordinary benefit

Fig. 34.

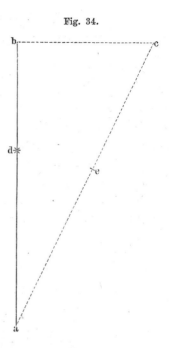

Figure 34. Line divided according to the golden ratio.
$bc = \frac{1}{2} ab$, $ce = bc$, ea conveyed to ab yields d as the
point of division.

that came from having recognized it. Theodor Ludwig Wittstein claims
the expression originates in the Middle Ages, appearing then as *sectio
aurea* or *sectio divina*, an honorific indicating the excellence of a theorem
that fascinated mathematicians of the time.[4]

Zeising begins his study with a historical overview, including an extensive summary of the work done by his predecessors on the theory of
proportion in the human body—including physiologists, artists, art con-
noisseurs, and philosophers. Among these are many familiar names, such
as Polykleitos, Lyssipos, Leonardo da Vinci, Michelangelo, Albrecht Dürer,
Raphael, Lavater, Horace Vernet, Johann Joachim Winckelmann, Johann
Gottfried Schadow, Adolphe Quetelet, and others. Zeising devotes particular attention to his immediate predecessors, history painter Carl Schmidt
and physiologist Carl Gustav Carus, the former for his work *The Key to
Proportion: A New System of Ratios in the Human Body, for Visual Artists,
Anatomists, and Friends of Natural Science*, the latter for his work *The The-
ory of Proportion in the Human Figure*.[5]

Recently, figures like Fechner, Conrad Hermann, Wittstein, Johannes
Bochenek, and Caspari have begun to elaborate Zeising's work.

According to Fechner, most natural scientists have remained more or less ignorant of Zeising's efforts, though he cites a few notable exceptions: Emil Harless takes Zeising's theory into account in his textbook on sculptural anatomy, as does Friedrich Wilhelm Hagen, whose detailed measurements confirm the theorem of the golden ratio in the structure of the head and brain. Furthermore, a number of Zeising's papers have appeared in scientific journals, and his views have caused a sensation in the field of aesthetics and have for the most part been met with assent.

Staunch supporters of his theory attach even greater value to it, and rightly so, ever since the acclaimed founder of "physiological psychology" subjected it to scrutiny as meticulous as it was impartial. Fechner makes no effort to conceal his qualms about the law's unrestricted application beyond the sphere of aesthetics; he also credits Zeising for properly grounding the golden ratio philosophically; he finds adequate support for Zeising's view of dimensional ratios but not of divisional ratios and admonishes that too much is made of the aesthetic advantage in the golden ratio's symmetry. Finally, Fechner declares that he is far from underestimating the merit and the meaning that is thought to have been discovered in the golden ratio, to a certain point. In fact, he considers it the first real discovery that has ever been made in the field of aesthetics and that it will have secured a place for Zeising's name in the history of the discipline. It is easier, he writes, to verify, restrict, and refine a discovery than it is to make one.[6]

Conrad Hermann begins his treatment of "The Law of Aesthetic Harmony and the Rule of the Golden Ratio" with an appeal to Fechner's judgment: "Fechner recently submitted these investigations and their findings to a thorough and essentially corroborative examination. It now in fact appears that even in the sphere of the beautiful something has at last been positively identified, in a truly exact or empirical manner. . . . We believe, in connection with Fechner's remarks, that we may now regard as fact that the principle of the golden ratio is of paramount importance for the inner structure or the proportions of the beautiful."[7] Hermann's paper has the merit of having advanced our appreciation of the golden ratio on the basis of a wealth of new observations that touch profoundly on the aesthetic question. He, unlike Fechner, is far from disallowing the golden ratio's application beyond the sphere of art, as Zeising would have extended it, to natural phenomena and even to ethical and religious concerns. Yet, while Hermann is not lacking for worthwhile ideas when it comes to the broader application of the theory, he nevertheless defends the primary task of observational aesthetics—which he sees as comprehending the immanent order of things—from being encroached upon by an arithmetical order, in a superficial-mechanical sense of the term.

Wittstein, in his paper on "The Golden Ratio and Its Application in Art," narrows the scope of his study—in contrast to the wide range of Zeising's—to a smaller field, therefore easier to survey, meanwhile bringing his own long years of experience to play in ultimately confirming Zeising's basic ideas. From the standpoint of immediate experience, though always with a view to the ideal side of the object, Wittstein proceeds with measured and certain steps to lay the groundwork first for a definition of the beautiful in general—which is fascinating in and of itself—and then thereby for the law of proportion of formal beauty in particular. Where others have simply assumed this law as given, Wittstein reveals its provenance and thereby also its future. When compared with the expected staying power of this work, earlier attempts of similar aim can be of merely historical value, though, as forerunners, they nevertheless have helped keep interest in the question alive.

Oddly, Bochenek does not even mention Wittstein's study in his own work on *The Standard Male and Female Figure in Accordance with a New System*.[8] We cannot assume Bochenek was unacquainted with Wittstein's work. Their separate studies are complementary. They are both based on the grounds prepared for them by Zeising, and though each deals with a different aspect of a common problem, they coincide in that each adduces substantial new material by which to expand the theoretical construct that both are assembling.

Bochenek is convinced that he has rediscovered the irrefutable laws of the beautiful which the ancient Greeks knew but have been lost to us. After an investigation into the basic forms on which rest all that is sensibly apparent in nature, art, and science, Bochenek concludes: "Recombinations of these basic forms create the innumerable structures that, upon scrutiny, resolve once again into these basic forms. All the forms occurring in nature, from the most primitive to the most complex, are united in the human figure. . . . And we therefore designate as beautiful not that which is composed of more or less forms, but that which is composed exclusively of forms that are vital to its purpose. Indeed, it is on this that formal beauty rests." Now, although Zeising does use the golden ratio to measure the proportions of the human figure, he does not represent the speculative condition, nor do his measurements lead to any conclusive purpose. Bochenek's task is to establish both of these things. At the same time, he acknowledges that Zeising's views on the system of proportions are the most profound that have yet been documented and that there is more truth in his exposition than in all systems combined, which he cannot but confirm through his own approach to the topic.

Bochenek thus goes beyond mere specification of proportion in the organization of the human figure, which was typical of other systems prior to his. In so doing, he avoids the cliff upon which his predecessors more or less foundered, in the opinion of strict aestheticians. Moreover, since there is no numerical ratio that perfectly corresponds with the golden ratio, Bochenek maintains that real organic proportions must be removed from all numerically determined metrics; what he has in mind instead is the ratio of ratios, as it initially appears as the symmetry in the body. "A line partitioned according to the golden ratio produces smaller, interrelated segments such that, in a sense, one segment represents the entire line, because it is a harmonic segment of the line's entire length."

The golden ratio is therefore an infinitely self-reiterating proportion, the most perfect expression of which is the human figure—except that this figure demands, as organism, the concept of "organization" rather than that of "partitioning." Parts lend themselves to quantification, but limbs, as organs, can only be grasped as the invariably constant ratio of variable dimensions. Nothing has been quite so detrimental to the theory of the golden ratio as the assumption that works of art could be created according to numerical specifications and a doctrinaire canon of beautiful proportions.

Wittstein too safeguards the theorem of the golden ratio against misunderstandings like these through a proper appreciation of the creative work of the artist—who "has the absolute right to refuse to accept anything that, as a rule, imposes itself upon him from without. . . . By contrast, it is fascinating to us, and also indicative of the philosophical urge inherent in human beings to seek out the causes of things, when we look back after the fact to discover the rules the artist must have unconsciously obeyed and when we then collect these rules into a theory of art. . . . Even the musician is unaware why exactly he has chosen to use as consonance the octave rather than the fifth; he does it because it pleases him, and then the physicist comes along after the fact and locates the reason for this in the tone's period of oscillation."[9]

The intuitive creative work of the artist and the "after-the-factness" of scientific explanation—indeed, these are the very essence of organ projection!—could hardly be expressed any more straightforwardly and impartially. The law of proportion has emerged as a recognized organic principle only by means of the artifactual objects that attest to it in retrospect.

It is this sense we should bear in mind, regarding the distinction between fields of art and artistic individuality itself, when we encounter, for instance, "the beautiful artist, in whom nature created art" in Robert Keil's recently published sketch on the life of Corona Schröter.[10] Wilhelm von Kügelgen intends the same thing when he comments in his *Three Lectures on Art*: "The artist initially represents nothing other than himself."[11] And Schiller, too, writes with respect to his *dramatis personae*: "As a plain white beam of light radiates thousands upon thousands of colors simply by glancing off a surface, so I am inclined to believe that within our soul all our characters sleep, each according to its primary substance, and that these acquire a manifest existence through contact with nature and reality or through artistic illusion. *All that is birthed of fantasy is in the end only ourselves.*"[12] And if we add to this an objective fact noted by David Friedrich Strauss—"Schiller could no more have written *Hermann and Dorothea* than Goethe could have written *Song of the Bell*"—then we find Ferdinand LaSalle's germane assertion, that absolute self-production is the most profound point of the human being, well corroborated not only by philosophy but by the performing arts as well.[13]

The human being's most profound point, it turns out, is the primal homeland of art. But since absolute self-production coincides with what we, in the interest of depicting the real cultural-genetic factors at play, have called organ projection, we will now have to discuss Zeising's fundamental law in more detail, in the context of organ projection specifically.

More often than not, attempts to develop a theory of proportion have attended too little to the human body as a whole and focused too much on the skeleton alone—as did Carus, for instance. Though it is inarguable that the skeletal structure provides the body its kinematic support, it is no less obvious that a singular focus on the fixed interior framework will err if it fails to account for the musculature that rounds out the body's exterior. The body thus divided exists nowhere but the imagination, and if only the anatomically prepared skeleton is present at the investigation, the living proportion has vanished.

The whole human being is figure, the skeleton itself merely a frame [*Gestell*]. The skeletal framework and the muscular system constitute one another and together amount to the body's "constitution." Georg Hermann von Meyer describes this interdependency in detail in his *Statics and Mechanics of the Human Skeletal Structure,* in which he writes: "If the parts of the skeletal framework and the whole framework itself determine muscle activity and the body's static proportions, then on the other hand we also see that all these relations have an influence on the configuration of the individual bones and of the skeletal framework as a whole."[14]

Given this succession of structures—membranes, vessels, nerves, muscle fibers, sinew, cartilage, spongiose and compact bone—merging into and emerging from one another, the line dividing the skeletal framework from the soft tissues dissolves.

Needless to say, this is why Zeising engages the *whole of the human figure* in his fundamental law of proportionality. The whole figure is what delimits the sculptural ideal in the realm of the beautiful in art. For such a law to exist at all, it must, as an emanation from the human figure, be immanent in the same; and Zeising also had to prove that while the law of proportion may not alone determine the beautiful in art, it is certainly one of beauty's several qualities.

At this point, Zeising goes on to develop his system with a discussion of the relationship between proportionality and beauty in general, and this is followed up by an explanation of that relationship in the spheres of art and nature, respectively.

In his section on the importance of proportionality in the realm of formal beauty, Zeising deals first with infinity and unity in formal terms, then with the harmony of both. This harmony reveals itself in three different stages: as *regularity* or *symmetry,* as *proportionality,* and as *expression* or *character.*

If symmetry appears primarily on the outside or in the outline of the figure, then proportionality proceeds outward from within it, from the cardinal point internal to the figure that lends it the character of unity. This cardinal point is the actual core or germinal point of the whole.

Expressive or characteristic beauty receives its laws predominantly from the domain of psychic motion—motion that "not only permits but actually commands a departure from the laws of symmetry as well as the free modification of the laws of proportion." Such departures are, however, not to exceed the point beyond which these laws would cease to facilitate said free figuration.

The law of proportion thus acquires the following meaning: the authentically beautiful expressive form is that in which the law of proportion remains recognizably expressed despite all modifications.

Zeising then explains the emergence of a proportional figure in the following manner: "Figures are formed through expansion of each germinal point into a number of different radial lines; these lines then expand further in width and breadth; ultimately, all of these lines link up among themselves to unite, as limbs, into a whole. . . . In this case, the line that forms the perimeter of the figure does not appear as the essential or original element of the form. Rather, the lines of a radial character that form the framework for the outline are this element, essentially the figure's so-called axes. These axes, however, tend to remain invisible, though they already also form real bodies, fibers, tubes, bones, veins, and so on.

"The ratio between the whole and the limbs is therefore the same as that which relates the limbs to one another. The outlines of the figure thus appear, with respect to these inner lineaments, as their clothing, so to speak—or, better: we should think of them as the threads that weave the outermost ends of the inner lineaments together into a self-contained whole.

"From the nature of proportion thus develops the character of progression and, with it, the character of growth and of organic life.

"With more complex figures of this kind, the primary tendency always appears to be *vertical*, or the dimension of height; the secondary tendency then is the horizontal, or the dimension of width. The remaining dimensions are to be considered only as mediatory between these."[15]

Bochenek further elaborates on this idea. He approaches the configuration of a creature in relation to a parallelogram, the right-angled sides of which enframe [*einrahmen*] the outermost length and width of the figure. Identifying this parallelogram as the *enclosure*, he assumes, and consistently confirms, that each creature is a formation of its enclosure. When observing formation in the animal kingdom, from the least creature to the human figure, one notices the longest axis spring up from a horizontal to a vertical position. No other animal resembles the human being in this respect; no other animal has a tall, vertical enclosure, nor does any other have such well-developed organs.

Bochenek's additional commentary here deserves our closer attention, since it touches curiously on the question of the theory of organic development: "The first and principal distinction between the animal and the human involves the elevation by degrees of a horizontal to a vertical line—or rather, the filling in of a right angle by various lines that may correspond with the primary axes of different creatures. Given that a seventy-degree angle cannot be said to emerge naturally from a fifty-degree angle, instead each angle must be regarded as an element in and of itself, then by analogy we cannot say that creatures' forms naturally emerge one from another; instead we must likewise regard each creature's configuration as an element in its own right."[16]

Each creature has its upper and under sides, but a distinction between the upper and lower body is warranted only in the case of the human form, owing both to the bisecting line of the central cavity and to the complementary groups of organs positioned on either side of it. Zeising points out that the organs belonging to the upper body are those in which the human being collects and abides in itself—the organs of nutrition, of the finer senses, and of reason. The lower body comprises those organs in which the human being divides and splits off from itself, giving itself over to others and to movement—the secretory and sexual organs and the organs of locomotion. On the one hand, then, the character of unity and of persistence within oneself; on the other, an image of diremption, schism, and letting go of oneself.

The zone of distinction between the upper and lower body is called the *girdle region*. It is formed of the cavity located between the bottommost ribs and the crest of the pelvic bone. The golden ratio's partition falls within range of this cavity, wherein it has some latitude, for even here "the law of nature permits variety, in order to ensure against a stereotypical uniformity."

The upper limit of this latitude is marked by the waist, "while the navel, which often lies somewhat lower—that is, slightly beneath the golden ratio—represents this latitude's center of gravity." Figure 35, taken from Ignaz Küpper's fascinating monograph *The Apoxyomenos of Lyssipos*, perfectly conforms with the rule of the golden ratio, as the scale beside it illustrates. The principal segments of the head, the trunk, the thigh, and the lower leg lie within the standard intersections, while point d intersects the larynx, e the bottom of the knee, and f the front of the ankle.

The navel—the nutritional portal of the fetus in utero, which terminates in the newborn's umbilical scar—is the emblem of the organs' commencing to function, the birthmark of the individual's connection with the universal. Hereafter, the navel "appears as the core and point of departure of the two unequal yet proportional parts, as the locus of proportional organization, as the human body's golden ratio—that is, the shorter upper body (from the navel to the crown) is to the longer lower body (from the navel to the sole of the foot) as the lower body is to the body's entire length."[17]

For those who remain unconvinced by the careful methods Zeising applied in discovering the standard proportions of the human figure, as expressed in the golden ratio, Bochenek's further demonstration—carried out in Zeising's own spirit and which, on the whole, ought to put the matter to rest—should suffice to eliminate any doubt that remains.

Bochenek pursues his aim with utmost rigor. Because he believes the human figure is formed of its enclosure, it is within this encompassing form that the fundamental ratio appears to him *a priori*. This ratio, endlessly reiterating within the enclosure, is receptive to a myriad of diverse, well-proportioned organizations. Bochenek does not think it advisable to support his argument by means of examples; rather, by attending only to that which immediately pertains to it, the object is supposed to bring forth proof on its own. "This way indeed," as he writes at the end of his essay, "we would find compelling proof of the lawful, harmonic development of the one from the other and thereby confirm that every other measure lacks all scientific basis."[18]

And so it seems the standard proportion, which from time immemorial has been the starting premise for so many brilliant explorations in both the arts and sciences, is not a phantom after all!

Now that it has been established, it will help to sustain and corroborate the ancient truth that man is the measure of all things and therefore also prove to be a powerful support for the theory of organ projection.

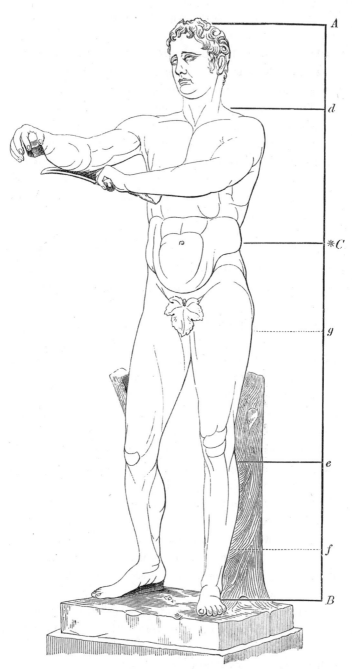

Fig. 35.

Figure 35. Master figure.

The entire sphere of the outside world is now called on to testify to the meaning of the standard proportion. There are in fact no exceptions where symmetry is concerned, and if something appears exceptional now, later research will prove the exception to have been illusory.

Zeising too was convinced of this and did not hesitate in his ground-breaking book to test this particular depiction of the law of proportion in the different domains of art and nature.

After examining the organization of the human body in terms of its lawful proportions and deviations from these on the basis of gender, age, nationality, and individual character, he goes on to investigate manifestations of the law of proportion in the context of other natural phenomena, both macro- and microcosmic (minerals, plants, animals). Then, toward the end, he dwells a while on its manifestation in the field of architecture and on its significance for the fields of music, poetry, science, ethical concerns, and religion.

It was of course inevitable that, as an individual, he fell short of providing a comprehensive solution to such an enormous problem, and eventually he had to concede that he had failed to provide evidence for many of his claims.

This, however, is the fate of all innovators—that errors and deficiencies in the details are unavoidable. Their successors' task, then, should the original premise prove inviolable, is to improve and to supplement, with moderate restrictions in the division of this work.

For us, the golden ratio, as the standard organic proportion, is valid beyond doubt. In it, *measure* [*Maß*], *proportion* [*Gleichmaß*], and *symmetry* [*Ebenmaß*] are unified.[19] Therefore, as the basis of harmony, the golden ratio is a quality of the beautiful, an ideal exemplified in the masterworks of Hellenistic sculpture.

We are making a strict distinction here between measure [*Maß*] and metric [*Maßstab*]. Measure is at work in the sphere of the organic; metric is inserted ready-made into the mechanism. Measure is the reflection of a relation among orders of magnitude, while metric is the expression of a number. In measure, life processes are in motion, while metric is enforced from without.

If the golden ratio is understood as a measure of relation among limbs [*Glied-Maße*], thus as a disciplinary measure [*maßregelend*], then Zeising is absolutely correct. Custodians of the mechanical outlook, however, are absolutely wrong in refusing to admit anything that is not a quantitative metric.

Measure, as relation, is without number, and no organ comes with an arithmetic label attached. What is a finger on its own? A decomposing thing, soon to expire. But think of the many primitive stone axes that remain to us: each one is still the same as it was millennia ago. There is no finger without the hand. On the hand, the number five amounts to an indivisible whole, indivisible like the colors of a spectrum or the poles of a magnet.

The binary quality in the fundamental organic ratio is also indivisible, in like manner. This ratio distinguishes a larger portion from a smaller, and yet the two together—the whole—is something other than either portion is on its own. Only two, and yet—three!

Now would be as good a time as any to point out that there is an air of the so-called *mystical* that lingers on in the term "the golden ratio." This circumstance explains in part why the peculiar authority of a simple geometrical theorem dating back to the medieval period has in modern times been met with a certain reactive suspicion.

Philosophy has meanwhile proceeded to cleanse the sphere of the mystical from the morbid excesses that once distorted it, while at the same time and to the same degree, advancing in the work of making the content of intellectual intuition accessible to rational demonstration. Indeed, Eduard von Hartmann perceives "in the entire history of philosophy nothing other than the translation into a rational system of a content mystically generated out of the form of the image or of unsubstantiated claims."[20]

We owe it to certain recent works—*Meister Eckhart, the Mystic* by Adolf Lasson, for instance, and von Hartmann's chapter on "The Unconscious in Mysticism" in his *Philosophy of the Unconscious*—that one no longer balks at the idea that the mystical may have some significance for the general development of culture. That said, we may as well rid ourselves now of any lingering bias against the golden ratio.

No one disputes the fact that we are surrounded on all sides by mystery and the inscrutable, that we ourselves are suffused with it. But if this mystery were to remain, without change, the content of an obscure feeling, then morbid stagnation sets in, and this stagnation exercises a chilling effect on all who are in the habit of confronting the uncomprehended with thought. However far we may advance in providing rational evidence for this mystical content, that is a concern all its own. But consciousness, merely by virtue of its wish to be able to and its attempts to do so, alleviates this stagnation, by placing the object within the stream of thought and in this way forever safeguarding it from the dangers of dogmatic impasse.

To be sure, even as philosophy clears the path ahead of itself, new riddles spring up behind it. In this way, material is constantly being regenerated, even at the highest levels of research, and fresh sustenance is being provided for our further encroachment upon the uncomprehended. Accordingly, each "wish to be able" is accompanied by a host of new questions. Indeed, each proof, each explanation requires its own explanation in turn—proof enough that humankind is sufficiently secured against a situation in which there remains nothing left to perceive or explain.

One therefore has to resist, not flee mysticism, not stamp it out entirely—least of all where what is at issue are the body's fixtures and processes, for these are processes and fixtures with which weights and metrics, with which numbers in general, cannot come to grips. What we are talking about is the living human being. In him, the two principal phenomena—body and consciousness, matter and spirit—are absolutely immanent, one within the other in his *being-as-organism*, a *being-one*. In this concept of being-one, "life and limb" are so beautifully and truthfully coupled as to have met in the very rudiments of language.

If the authentic work of art is already incommensurable, then how much more so is that most sublime prototypal image of the beautiful in art—the "living, breathing" human being? While objects of technology are amenable to being precisely measured and counted, the organs and their functions are entirely recalcitrant by contrast. And yet, though it may occasionally appear otherwise, because life is ceaseless change and transformation, the results of any measurement will differ from moment to moment and therefore never be exact.

When the dimensions of a statue, then, are submitted to a metric, it is the marble, or whatever the statue may be made of, that is actually being measured, as any other material would be. But the artist's idea which is realized in the material stands remote from number. The idea in the artwork corresponds to ensoulment in the sphere of the organic.

The artwork's encounter with exact specification is but an expedient, borrowed from mechanics, just as we find borrowed mechanical expressions in our attempts at understanding organic formation. What we perceive of number in artworks and organic forms defies all measuring devices.

Numerical ratios therefore cannot coincide with the one proportion that reiterates itself endlessly in varying dimensions within the golden ratio. Any attempt at applying such unstable numerical expressions to human beings or to works of art is bound to fail. For human beings and works of art are the only two real existences, according to Conrad Hermann, "because each of them is, in like degree, of a sensible and a spiritual nature. For this reason, they appear as if standing on the boundary between the two general regions, or spheres of all being—the real and the ideal."[21]

In the event, in the context of art, that one resorts to using a numerical ratio that approximates the golden ratio as nearly as possible, either this plays a role in teaching the rudiments of technical composition and in training the eye to detect symmetry and proportion, or it serves as a visible test for theoretical agreement. Zeising then offers a surprising corroboration of his own theory in a number of renderings of famous works of sculpture. In so doing, he allows the works of art, in which artistic genius has unconsciously brought forth a beautiful symmetry, to attest themselves to the validity of the retrospectively identifiable standard proportions in the human figure.

According to Zeising, "the human body is an organism sprung forth from an original idea, organized according to a fundamental ratio, imbued with perfect harmony and eurythmy," and "it is in the human figure that a fundamental principle is first and most perfectly realized—the fundamental principle of all formation, a principle compelled toward beauty and totality in the realm of nature as well as art."[22] It follows from this that the fundamental ratio must prove its constitutive power—a power that extends even as far as the human need to give form.

Having acknowledged the rich yield of Zeising's research in aesthetics, insofar as fine art is concerned, we will now turn our attention to the precursory stages of three-dimensional artistic representation—that is, to *handicraft* and to the *applied arts,* which is to say, to that domain with which our own investigation is primarily concerned.

Recalling that the extremities of the human body become limbs of measure through their permutation into norm and number; that these measures, borrowed from the body, are primarily the simple dimensions outwardly evident in the arm and hand and foot; and that, under the assumption of a general principle of organization that regulates specific forms, the length of the foot and the number of the fingers are not incidental but institutive laws of the body—keeping all this in mind, we will pause to examine the organ from which handicraft derives its name, as well as the entirety of the arm immediately attached to it (Figures 36 and 37).[23]

The first partition of the arm outstretched from the shoulder to the tip of the middle finger exactly intersects the crease formed on the inside of the elbow joint, or the narrowest point between the upper arm and forearm. The upper arm (from the shoulder to the inner elbow) is to the longer forearm as the forearm is to the arm's entire length. This same ratio is formed when the upper arm is partitioned at the armpit and when the forearm is partitioned at the wrist. The ratio is reiterated several times in the organization of the hand, which has the distinction of being, beside the head, the most intricately constructed limb of the human body. Zeising and Bochenek both include a number of illustrations showing how these ratios in the arm correspond not only among themselves but also with those of the torso beside it and furthermore with the ratios evident in the body's total organization, both width- and lengthwise.

Since these ratios, as Zeising notes, appear most perfectly constructed in the head, Figure 38 is included for the sake of comparison. Copied from photographs of Alexander Trippel's bust of Goethe, which one can purchase at the library in Weimar, this illustration provides a wonderfully characteristic example of the standard proportion.

The muscular system as well as the body's other interior structures are also subject, in form, location, and arrangement, to one and the same standard ratio—only it is much less apparent in these than it is in the skeletal structure, which again brings up the question as to whether the brain cambers the skull or vice versa. All the same, there appears a combination proceeding to infinity, mocking all numeration, a combination that, containing measure and limit within itself, "signals in the golden ratio the unification of the ultimate distinction—that is, the distinction between unity and diversity."[24]

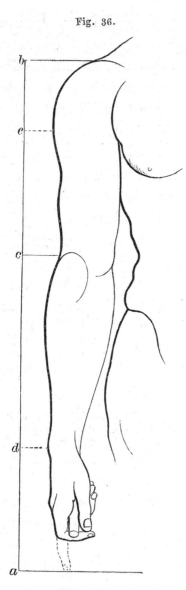

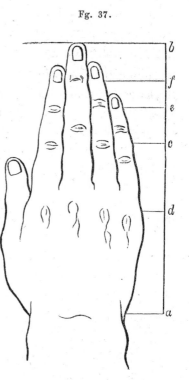

Fig. 36.

Fg. 37.

Figure 36. The arm.
b height of the shoulder, *a* tip of the middle finger, *e* height of the armpit, *d* wrist.

Figure 37. The hand.
ad the back of the hand from the crease of the wrist joint to the knuckles; *db* the entire forehand; *dc* the base of the fingers up to the creases of the knuckles in the index and ring fingers; *eb* the ends of the fingers up to the tip of the middle finger. Thus *ad : db = db : ab; dc : cb = cb : db* and so on.

Proponents of Zeising's theory who venture to claim for the standard ratio a ubiquity throughout the entire human body, determining even the ramification of the veins and nerves, are nonetheless modest enough to wait for advances in relevant areas of research to provide the desired evidence, beyond what has already been objectively established in physiology. There is no shortage of promising attempts these days to dismantle the already rather weakened barrier between physiology and psychology, a barrior once so laboriously defended. These attempts betoken the

Fig. 38.

Figure 38. Pattern of a head.
cb upper part of the head, from the crown to the orbital rim: section *d* indicates the beginning of the hairline; *ca* lower part of the head down to the middle of the neck (larynx); *e* the base of the nose; *f* the jut of the chin.

eternal authority of the fundamental ratio, as the organizing power enfleshed in the human figure.

By now we understand the sculptural artwork's gesture toward the
living human being, and so at this point we find ourselves on a path that
will take us directly into the craftsman's workshop. The "gilded ground
of the handicrafts" and the "golden ratio"—why should their adjacency
be symbolic rather than speak to a matter of fact?

What we need to consider now is the question as to what right a
minority of primitive tools have to be designated as hand tools in the
superlative sense, in light of the particular importance that the quality
of handiness has for precisely those tools without which human culture
itself can hardly be imagined. A tool is always easier and more comfort-
able to handle the greater is the influence of the organic standard ratio
on its form. No matter whether this influence is felt as a factor of usefulness in the sphere of the handicrafts or as one of pleasure in the industry
of applied arts, it is always the golden ratio that, uniting use with beauty,
promotes the useful by means of the pleasing and beauty by means of the
serviceable; it is the golden ratio that, set on the border between atelier
and workshop, assigns a rank and place to each piece of equipment in the
system of human needs.

Our belief that the first tool to proceed from the human hand was
the actual impetus for the development of culture does not engage with
whatever assertions the anthropological sciences may make concerning
the simultaneity or priority of word and idea or of language and tool,
given the limits that we have imposed on our investigation.

Archaeology, which axiomatically assumes the human being's original
disposition to culture, grasps the primitive human being in exactly the
right way—when it takes him by the hand and locates in the first ax or
hammer the first real human deed. The human being is in fact the "tool-
making animal," according to Benjamin Franklin. And now that we have
the human being at hand, or in hand, as it were, it makes absolutely no
difference, with respect to further discussion of the fundamental ratio,
whether the route we follow in our exploration of the hand ascends from
the fingertips, over the hand, arm, and torso to the section of the ratio
represented in the standard length of the body, or whether our exploration deposits us, via a line descending from the body's entire length, back
at the hand and the creases of the knuckles.

No matter where on the human body we measure these basic dimensions, what we are recording, in the unity of the ratio, is each of its distinctions and, in each of these distinctions, the unity of the body. Bochenek explains the method by which these measures may be taken, without any arithmetic, using only an adjustable or reduction compass specifically calibrated for this purpose.

"If one were to place the longer leg of the compass on a line divided into eight equal segments, then the shorter leg would have to encompass not quite five of these segments.

"Then draw a straight line, take the same line with the longer leg of the compass, and position the smaller leg of the compass on it to get a short and a long segment. If this long segment is plotted with the longer leg, the short leg must likewise agree with the shorter segment. As soon as there is no difference between these segments the compass is normally calibrated.

"Without this ratio, it is impossible to follow my system comparatively. It is therefore advisable to equip oneself with such a compass. Other methods that use a substitute for the compass are much too cumbersome."[25]

In the event an adjustable compass is not available, Figure 39 provides a useful scheme for crafting a device of any size with which to measure proportions. If a line *ab* is divided by *c*, in accordance with the golden ratio (see Figure 34), and if a point *d* is placed at any distance from *ba* and the lines *db, dc,* and *da* are drawn, then any line drawn between *db* and *da* will be intersected by *dc* in the proportion of the golden ratio.

If only in order to provide, where possible, clear numerical evidence of his theory's unfailing accuracy to those who may be skeptical of the proof that we have found preserved in marble, Zeising was accommodating enough to have applied a procedure that, because mechanical, runs counter to the concept of standard proportion.

One has to consider this courtesy little more than an expedient, if one also takes seriously Zeising's express protest "that the law can never quite be reached through finite numbers, that it is therefore both measurable and incalculable, both rational and irrational, extremely clear at the same time as it is clothed in the allure of a depth that will never quite be fathomed."[26]

At any rate, application of Bochenek's procedure should mitigate these and similar misunderstandings. Still, one should always keep in mind

Fig. 39.

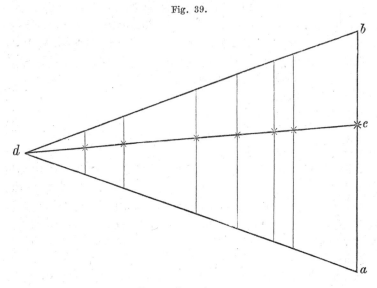

Figure 39. Proportion meter.

Zeising's caveat concerning the existing latitude, wherein the individual phenomenon is left to come to its own terms with the law.

Expecting comprehensive pursuit of the law to provide evidence of, on the one hand, how the world-creative power accomplished the most extraordinary effects with the most apparently meager means and, on the other, how the passage from the one into the infinitely many and diverse can possibly have taken place, Zeising goes on:

"This can be achieved only if each scientific discipline submits the law to a rigorous examination that is specific to its own particular standpoint and then compares the results of its observation and experience with the provisions pursuant to the law itself.

"Of course, one can neither expect nor demand that individual real phenomena agree perfectly with the law, because each individual phenomenon is as such necessarily imperfect to a certain degree and cannot therefore correspond to the law in each and every respect. Indeed, a phenomenon may acquire the *appearance* of perfection simply by breaking away in some degree from the law of the whole, in this way giving its particularity and contingency the aura of a unique totality and freedom.

"The law should therefore be considered only as an *ideal type* or *norm* which actual formations approximate more or less closely. In the sphere of individual phenomena, we should by no means regard those formations in which the law is actualized with rigor and inflexibility as its perfect realizations; rather, the law is most perfectly realized in those formations that, like the human body, possess the full power of inner life and self-determination, through which the law appears as if sublated [*aufgehoben*] though in fact it is only set in motion and in flux and may be perceived in ways both higher and less restricted."

In support of this, Zeising quotes from the classic passage in Goethe's essay on Geoffroy Saint-Hilaire's *Principles of Zoological Philosophy*: "Should we continually see only what is regular, then we must think it must be so, thus determined from the start and therefore fixed and immobile. Should we see, however, the deviations, malformations, the horrible disfigurations, then we recognize that the rule is fixed and eternal and yet at the same time alive; that the being does not emerge from the rule but within the rule can reshape even misshape itself; at all times, however, as if bridled, the ineluctable dominion of the law must be acknowledged."[27]

Both the requirement of and the license within that free latitude are, for handicraft and the applied arts, vindicated accordingly. The degree of handiness or symmetry, whether greater or lesser, is not independent of the individual phenomenon's scrupulous agreement with the standard proportion; but, more importantly, handiness and symmetry must comply with the latitude that guarantees mobility, a latitude that is to be found in the space between two thresholds, the one of too much likeness, the other of too much unlikeness.

We will now focus on a familiar and exemplary hand tool, in order to provide evidence of the above.

I once looked on an as old backwoodsman in West Texas demonstrated what he called the *philosophy of the ax*. Placing his American ax beside one manufactured in Germany, he explained in rough terms the differences between them: the shaft of the German ax was straight as a post, rigid and refractory, while the slender "handle" of the American ax displayed a pleasing longitudinal undulation; the iron head of the German ax was attached at a rigid right angle, while the American ax head angled slightly inward and downward. He then dwelled on the latter with unusual appreciation, running his calloused hand over each individual swell

Fig. 40.

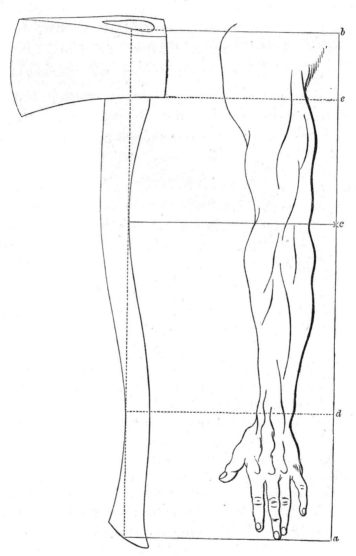

Figure 40. American ax and the human arm.

Fig. 41.

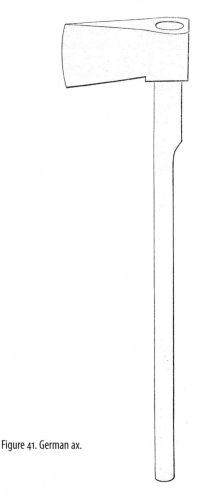

Figure 41. German ax.

produced in the iron as it was forged. He explained the ax's advantages, in terms of a powerful forward strike as well as an easy, relaxed backswing. He remarked on the qualities of the wood used to make the handle—a tough hickory, yet pliant and elastic enough to deflect bounce and protect the hand. That was evidence enough for this novice, and certainly a tangible impetus to further reflection! The obvious question now is, why is it that this ax can accomplish two or three times as much in a single day's work than the German ax can (Figures 40 and 41)?

A look at the sample cards of axes produced by the Collins Company of Connecticut and by the Douglas Axe Manufacturing Company in Boston indicates the unusual care taken in the manufacture of a tool on which depended, more than on any other, the establishment of the first settlements. One will note that the different models are named for individual states of the Union and ordered with respect to the needs of different regions and professions, with respect to national habit and age: the Yankee, Kentucky, Ohio, Maine, Michigan, Georgia, Turpentine, Fire Engine, Spanish Labor, Boys' Handled Pattern, and so on. It strikes one that this is no different from the way renowned violin and piano manufacturing firms elsewhere advertise their own wares.

Given all the many different ways the ax is adapted to various purposes, its basic form invariably conforms with the standard proportions of the organ that it serves. There are certain implements that are impervious to substantial improvement. The American ax is one such implement. Its simple and appealing shape suffices to meet the demands of efficacy and practicability in each and every instance.

The curves of the handle are constructed such that, were one to draw its longitudinal diameter through the center point of both ends, at no point would this line fall outside the handle's volume. Now, if we designate the end of the shaft that is fixed in the iron head as the upper, the opposite end as the lower, and we imagine the length of the ax to be the same as that of a worker's arm, as is conventionally assumed, then the upper end corresponds to height of the shoulder and the lower to the tip of the middle finger of the outstretched arm. In this case, the elbow joint coincides with the middle of the first inward curve of the handle, located just below the head. The smaller, upper portion is to the lower, longer one as the longer portion is to the entire length of the ax. Within these two sections, the same ratio is reiterated in the same manner as described above with respect to the upper and forearm, at those points corresponding to the armpit and the wrist. The points of intersection corresponding with the elbow and the wrist are the same as those points intersecting the longitudinal diameter, on the upper end, at the outermost edge of the handle's top curve and, on the lower end, at the innermost edge of the bottom curve.

As far as the head is concerned, the findings will be similar to those yielded by the compass method, though at first glance they may be somewhat less obvious. Since, in smaller objects, the relations among length, width, and depth overlap more because more condensed, precise specification becomes more complicated, and frequently these specifications fall within the scope of the latitude permitted within the ratio. A tool like the American ax is perfectly finished to the extent that its formal relations are a perfect reflection of those of its organic prototypal image. This agreement affects the symmetry that imparts practicability to the object and lifts it, beyond the useful, to the sphere of the pleasing, therefore to the realm of the applied arts.

What we've said about this one hand tool of course applies more or less to all the others. Its usefulness is determined by the degree of its handiness, this in turn by the measure of its consonance with the fundamental ratio in the body's articulation, a ratio that is unconsciously carried over into the tool.

This consonance is at once a harmonizing of the tool according with the fundamental organic ratio as well as the tool's proper attunement with the organism, on the analogy of a musical instrument—only in this case, it is a coarse, material, *artifactual* conjunction [*Fügung*].

When considered etymologically, the word *artifact*—borrowed from Latin, meaning "creation of the human hand"—reveals to us a deeper significance, to which we owe a new insight into the nature of the standard proportion.

The expression points back to a series of word forms deriving from the ancient Greek radical ἄρω ("füge" in German)—ἄρτι, ἄρτιος, ἀρτύω, ἀρτύω, ἄρθρον, ars, artus, articulus—through which shines a principle of symmetry common to both nature and art.[28] Since there is not space enough here to discuss all that this sequence of derivations may imply linguistically, we will limit ourselves to the one that is most relevant to our discussion.

Until only recently, the prevailing opinion held that in Pythagorean number theory ἄρτιον meant "even" and περισσόν meant "odd." But Richard Hasenclever has shown, in his above-mentioned book, with reference to the ancients and to subsequent hints left by Galileo, that each word had quite another meaning within the academy, that in fact ἄρτιον was taken to mean a unity divisible unto infinity and that περισσόν then meant discrete unities which could be multiplied into larger quantities *ad infinitum*. "Now then, as Nicomachus said, if the unitary value is compared with a living organism like a plant or an animal or with the entire universe, even the cosmos, then the unit can very well be called, in this sense, a ἄρτιον—that is, something *whole* or *complete*; and, in turn, quantities multiplicable beyond all conceivable limit may be called a περισσόν—that is, something *overmuch* or *outsize*.

"But the unlimited divisibility that is, so to speak, enwombed in the concept of the whole now has an exact expression in the sequence of aliquot fractions, while the quantities resulting from the infinite multiplication of units have theirs in the sequence of integers."[29]

Following on Albert von Thimus's study of *The Harmonic Symbolics of the Ancients*—which Martin Katzenberger has called a work of monumental importance—Hasenclever terms the former species of number the *divisible (τὸ ἄρτιον)* and the latter the *indivisible (τὺ περισσόν)*.[30] He then presents these in a graphic series, both in their reciprocal relation to one another as well as in their inseparable intersection *(ἐναλλάξ)*, thus:

$$\frac{1}{\infty} \ldots . \frac{1}{6}\frac{1}{5}\frac{1}{4}\frac{1}{3}\frac{1}{2} \quad 1 \quad \frac{2}{1}\frac{3}{1}\frac{4}{1}\frac{5}{1}\frac{6}{1} \ldots . \frac{\infty}{1}$$

"Proceeding from the unit (1) to the infinitely large $\frac{\infty}{1}$ on the one hand and to the infinitely small $\frac{1}{\infty}$ on the other, the integers (the indivisible) form a continuous *arithmetic* progression, while the aliquot fractions (the divisible) form a *harmonic* progression. The unit itself always forms the geometric mean of two equidistant members, which are always reciprocal values. Consequently, we find *arithmetic, harmonic,* and *geometric* proportion all within this natural numerical sequence—a conjunction [*Verbindung*] which is exceptionally important for harmony."[31]

It is recommended that the reader who is knowledgable about music and has an interest in ancient harmonic symbolics engage for himself with Hasenclever's commendable book in order to determine just how conclusive its arguments may be. So far as our purpose is concerned, we will keep only the ἄρτιον and attempt to relate this as closely as possible with the fundamental ratio in the conjunction of the limbs [*Gliedfügung*].

Doubtless, the terms "divisible" and "indivisible"—much more so than "even" and "odd"—are relevant to an understanding of music. But, generally speaking and more relevant to my own purposes, I find the terms *innumeration* and *enumeration* are more explicit. Innumeration is defined as a *number bound* within the unit as difference, while enumeration is defined as the sequence of discrete units, adjoined consecutively with one another in a continuous arithmetic progression, *the unbound number.*[32]

Take, for example, the hand, as the factotum of culture: the five-finger distinction is an innumeration, while a selection of five from any other set of objects would comprise an enumeration. The latter may be added to or subtracted from without the individual members of the sum or the remainder suffering any change; by contrast, the continuity of the innumeration is impervious to such a violence. The innumeration as such is inviolable; it is an accord—*ad cor*—in the living heart of the whole, which is to say it is guaranteed in the concept of organism, the whole of which is impaired by the perversion of any part.

The above-mentioned reciprocity in the sequence of the divisible and the indivisible is nonetheless not absent from the sphere we are dealing with here. For the innumerated distinction inherent in the unitary organ resolves into the enumerated external diversity of mechanical objects. The names for these enumerations—one, two, three, and so on—are words like any other words, elements of language spoken aloud, but in their mute abbreviated form, as numerals or ciphers, they become the ethereal tools helping to orient the human being in the world and in himself.

We also concede that numbers are pure representations and that in nature, in what is external, there are no numbers, only numerable things, objects that numbers may be applied to.[33] Since numbers are not originally derived from external objects but instead emerge from the mysterious depth of the fundamental ratio in the body's organization—which is the universal fountainhead of our knowledge and ability—numbers must actually therefore originate in an inborn organic distinction. The representation of this distinction, projected outward from within, could then be transformed into the reality of numerability and the enumeration of individual things.

Ethnographic observation has allowed us to conclude that the primitive human being's first attempts at counting were almost unbelievably difficult. The process is best described as follows: It did not occur to the human being from the outset that he might avail himself of the obvious, so to speak, fellow entities in his familiar natural surroundings to begin to develop a concept of number. On the contrary, it was only in the protracted and rather makeshift course of producing his primitive equipment that he was compelled to do so.

Enormous concentration was required of him in order to hold onto each means of alleviating his circumstance that came within his reach. The manner in which the most important and astounding inventions of the historical period came about hardly compares, in light of the considerable supply of resources already then available for use, with the tremendous expenditure of time and effort required in these first attempts.

The first hammer, the first bow! How many times the human being must have selected, tested, designated, altered, ruined, and begun again with his instruments! First there was *one*, then *another one*, and then *that one*; to each corresponded a thumb, an index finger, a middle finger—as many different names for discrete objects as one hand could signify. Only as manual dexterity improved did these first terms begin to transform into the names of numbers proper.

Even today there are still indigenous peoples who do not count higher than three. But once the number five had been achieved, it was only a small step to ten, from which the entire suite of the numbering system is thought to have advanced. By figuring out how to count to five on the fingers of one hand and to ten on the fingers of two, and, at the same time, by figuring out the number two, on the basis of having two hands but also on the more general basis of a consistent bilateral organization of the body that yields the difference between right and left, the ten fingers and two hands paired in the number twelve in turn yielded an even greater flexibility of division.

Since the names of the fingers as well as the designations right and left are meant to have played such a key role in the emergence of the words by which numbers are known, let us cite a passage from Lazarus Geiger that is concerned with the numeral *eight*.

"Can we justify expanding our explanation of the technical counting stratagem employed by so many primitive peoples, whereby the hand functioned as a rudimentary calculating machine, to explain the emergence of simple numerals themselves?"[34]

From an ethnopsychological standpoint, the answer can only be yes, especially since Geiger himself then adds: "Given the tremendous meaning—one we find it difficult to comprehend today—that the human body had for the intuition of the primeval world, it is right to assume that numerals were not originally intended to be related to the calculation of any object other than that of the fingers themselves. It was doubtless interesting enough for the human being to become cognizant of his five fingers as well as his two hands and two eyes. His interest in this knowledge, which he had first to discover, was occasion enough for him to have invented an expression specifically for counting it. From this point, he may have used the expression for counting other things, namely, those things which he noticed were present in numbers equal to the fingers on his hands."

In the section subtitled "Fragment," Geiger submits hand and foot, arm, and hand and finger to a comparative linguistic analysis. In the course of his discussion of quite diverse counting methods, he dwells primarily on the hand. Though its tremendous importance for human activity is hardly reflected in language, the hand is for Geiger the organic foundation from which the concept of number develops and, as such, cannot be overestimated.

His extensively detailed commentary here puts beyond a shadow of a doubt that the linguistic distinction among the fingers was the primary factor in the development of the first simple numerals.

The names borrowed from the "rudimentary calculating machine" remain constant in things! Counting by means of the organization of the fingers on the hand represents a tremendous breakthrough in the spiritual development of the primitive human being, for whom *the conceptual separation of limb numeration from limb differentiation* must have been just as difficult as the above-mentioned conceptual separation of force from motion.

The hand's natural grabbing after spoils and food, its grip on a tool, the mind and body's grasp of lower and higher technologies all help to liberate the universal-spiritual act of apprehension—the *concept*. The prehensile organ's disposition, perfected through practice, thus accommodates the very thought that urges to manifest phenomenally as idea, by virtue of an original, convivial artistic capacity.[35]

But the full power of such an evident reciprocity rests in the fundamental ratio, which enables our awareness of law and rule in the phenomenal world, which have their source in the superabundance of the standard human figure. For this ratio teaches one to conceive *innumeration and enumeration as principles of organic and of mechanical formations, respectively*.

As soon as the enumeration is applied to a specific purpose—say, in a self-contained machinal construction—then it reflects, as if in memory of its provenance, especially when converted into a mathematical formula, the organic conjugation [*Ineinsfügung*] from which it emerged.

If these mathematical formulae are objectively appreciated as works of art in light of their composition [*Gefüge*], it is because they gesture backward, along the path strewn with problems both physical and cosmic that they have succeeded in resolving, to the standard work of art in which they originate, the human being. In this same way, the human being becomes increasingly aware that his own hand is the universal organ of his ongoing self-revelation.

Without question, then, the hand and what it creates remain the real ground of human self-consciousness. Indeed, no matter how transcendental their flight, all art and science are never rid of their relation to the hand. The human being may pivot and turn all he likes, but his hand remains at play in all he thinks and does.

Just as the hand grasps and holds all things, it too is seized and sustained in a solemn contract, according to which it stands symbolically for the entire human being. And therefore science too, though this may not always be clear to it, has the historical human being by the hand at all times, and it must never loosen its grip, so long as it is willing to admit that the first tool is the start of historicity as such.

We too return time and again to the hand and to the assertion that the more perfect—that is, the handier—the tool is, the more purely its formal ratios align with the organ operating it.

The virgin forest testifies to what extent this is the case with the American ax. A less wieldly tool could never have overcome the challenge it presented to human fortitude. Without an ax like this, it is nearly impossible to imagine a culture advancing so rapidly across an entire continent, founding new states as it goes.

With respect to the seizure and occupancy of newly discovered territories, no one doubts the world-historical importance of gunpowder weaponry. Few, however, recognize the parallel importance of the American ax. And yet, it is equally responsible for the conquest of a hostile nature—a conquest accomplished only by virtue of the fact that the clearing of the native wood went hand in hand with the wiping out of the native people.

If one can imagine the sublime dimensions of the western continent and its vast virgin forests, out of which, in a sense, the ax carved new states, then one will understand the backwoodsman who speaks of the *philosophy of the ax*. One will also then be unable to avoid linking this expression with the great international scene that is on display in the New World and with the first ever competition under way on its own soil between a young centennial culture and the Old World that has existed for millennia.

The United States' supremacy over other civilized nations in the sphere of tool and machine technology is widely acknowledged. And its progress is swift in nearing the point in time at which it will have caught up with the older industrialized nations in the sphere of industrial production. What this rate of progress owes to the practicability of its tools, and to what extent the fundamental organic ratio has disciplined this practicability, should be evident, indeed literally so, in the word *manufact*.

On the one hand, rarely has an untamed nature of such immense size and promise demanded so much of the human being's inborn need to give form. On the other, never has a prototypal image as sublime as the industry of the Old World, promoted on all sides by art and science, come forward with such a wealth of forms to meet the human being's urge to formation!

As far as concerns household utensils in general, it would be both instructional and amusing were one to experiment with placing an adjustable compass on the various appliances and decorative objects in the kitchen, living room, and workshop—provided that a reasonable procedure were in place to keep from trivializing the experiment.

The principal ratio of the *handle* (grip, shaft, hilt) to the part of the tool directly responsible for changing the form and motion of the affected object should for the most part be immediately apparent, as should the ratio prevailing within each of the tool's principal parts. This also holds true for the size and shape of other everyday consumer and luxury goods, from the simple octavo notebook to different styles of dress, as Wittstein's study shows.

We have attempted to demonstrate, taking the tools of trade as an example, that the quality of handiness is but a product of the fundamental organic ratio. To this we now add: the artistic itself can have no other provenance than this—including the artistic element in technical craftsmanship. For, when it comes to artisanry, the practical and the proportional—these being synonymous—actually interpenetrate, and this in fact establishes the real basis for the claim that exact, superficial technologies can and do arouse pleasure.

The most precious vase would lack in beauty what its dimensions lack in proportionality, while the simplest craftsman's tool, provided it satisfies the standard proportion and its workmanship is pure, may be reckoned unconditionally pleasing.

Zeising writes at length of the importance of the golden ratio for every branch of technology "that may not consider the creation of the beautiful its ultimate or highest aim but that nevertheless regard the satisfaction of the aesthetic sense as a higher need, even though its products be intended first of all for use. Since these products cannot communicate the character of beauty to their works through submission to a higher ideal or by means of an expressive figure, all that remains as the sole means of exerting an aesthetic effect, once the object has been drained of its color, are its purely formal proportions.

"When contemplating objects like tables, chairs, cabinets, clocks, vases, bowls, pitchers, sconces, lamps, urns, and other household items and objects of commerce, or even purely ornamental objects like arabesques, rosettes, trims, ceiling decorations, wallpaper patterns, or even costumes, weaponry, toiletries, and so on—if one were to ask oneself, when they fail to please, what could be the reason for their unattractiveness, almost invariably the answer will be that they have violated some rule of proportionality, whether it be that height appears disproportionate to width, the dimensions of the parts disproportionate to the whole, the outward disproportionate to the inward-bending curves, the organization of one section disproportionate to another.

"So it is readily apparent how important it is, especially for the arts that have to reconcile necessity with beauty, that there be a reliable law of proportion on which they can depend. Moreover, we see how a tasteful and agreeable lifestyle is intimately related with the broad application of this law in these types of production."[36]

Doubtless Fechner had this passage in mind when he writes, in his essay *On Experimental Aesthetics*, that "greater benefit may be expected of objects with intrinsically pleasing proportions—objects having no claim, perhaps, to any higher meaning, but that were nevertheless intended to have an aesthetically pleasant effect. Among these are the works of so-called tectonics, which include not only the art of vessels, equipment, furniture, weapons, clothing, carpets, jewelry, and the like, but also architecture—which lacks no concern for higher ideals—and, ultimately, ornamentation."[37]

Of course, Fechner only further substantiates Zeising's theory in his latest work, *Primer in Aesthetics*, wherein he confirms and develops the materials provided by Wittstein's study.[38]

In response to possible objections that there exist a number of hand-held instruments that are nevertheless useful in spite of the fact that they don't obviously conform to the golden ratio, let us first and foremost say that these instruments would be entirely useless if they did not conform exactly with the organ that guides them and that, moreover, their usefulness actually increases in equal measure with their degree of proportional agreement. This is exactly the reason why the majority of hand tools have been so markedly improved in the context of an American praxis. What does the Yankee know of the golden ratio? His only concern was to minimize his stress in handling the instrument. But we look back and discover that he *unconsciously projected* his own fundamental organic ratio into what his hand constructed, in symmetry with himself.

Now, if we look within the sphere of the applied arts for the perfect example of a design emerging in the process of exactly this kind of organic projection, we will find none better than the *violin*. Julius Zöllner's reflections on the violin lead to findings that perfectly corroborate the above.

In his overview of the history of violin making, it seems to him as if the secret of proportion that earlier violin makers appear to have discovered *by virtue of a particular instinct* has since been lost and therefore that the inexpressible beauty of an Amati, a Guarneri, or a Stradivarius can only be achieved to a certain degree by imitating their design. Turning then to the component parts and theory of the violin, he remarks: "It is difficult to say which of the individual parts of the violin and of other similar stringed instruments are active in the production of a tone. There are so many assorted gradations and such fine nuance among them that one can hardly distinguish the unique influence that a particular part may exert in the assembled product.

"Savart attempted to develop a theory of the violin on the basis of physical principles, though without success, because the coffin-like instrument he assembled of six rectangular boards is absolutely incomparable with the violin. The laws of harmonic oscillation, derived from experimental physics, are made more complicated in the violin, first by the unique construction of its form, next by the camber of its sound-board, by the incision of its sound holes, by the variable thickness of the wood, by the fastening of its edges, by the sound posts and supports, by the variable distribution of tension along the strings, and so on. Although simple laws naturally govern each of these factors, the final outcome cannot be captured in a simple formula. This is also the case for the ribs, the back, and the neck. The action of each of these parts cannot be examined exhaustively in and of itself, and for this reason test apparatuses from which some part of the instrument has been strategically removed will never yield observations that lead to straightforward conclusions.

"But that is not to say that the physical sciences should abandon all effort to explain or give reasons for this instrument. On the contrary, their findings could benefit instrument makers, *but only if they start from the assumption that the instrument is a finished product and then attempt a posteriori to trace its uniqueness back to its grounds.*"[39]

According to Zöllner, because the violin is a product found by virtue of a particular instinct, it must be taken axiomatically as a finished product; the proportions of its construction are a mystery, and one must trace its uniqueness *a posteriori* back to its grounds.

Meanwhile, we know well enough that there is only one right trace to follow, and where it leads is back to the organism from which the discovery unconsciously emerged, as if by instinct. This "tracing"—what an unparalleled expression for the nature of organ projection! One cannot trace a phantom, because the phantom leaves no trace behind, while reality certainly leaves detectable traces of its becoming. Zöllner himself also quite clearly and beautifully points in the direction of the right trace, suggesting that:

"The violin, such as it is, is an inspirited instrument, an organism, having an animated nature as an organism does; it has a body, nerves, and a soul. Each of these depends on the others in a natural way, and none may be separated from the rest and its animating influence separately weighed and assessed."

Obviously, Zöllner means it figuratively when he calls the instrument an organism. He does so for the sake of providing visual coordinates for the traces, directing us toward the inner relation that binds the artificial construct with the human being that constructs it.

The violin is an absolutely complete, self-contained object. Analysis of it will succeed only by following the trace we have clearly articulated here. For the fundamental organic ratio is the master key to all its after-images, and these in turn are the copy keys allowing access to each of the organism's individual sensory entryways. On the one hand, the sound of the instrument will be dissonant in the hands of an unskilled player, and if something is wrong in its proportions even a skilled player will be out of tune. On the other hand, a mutual self-recognition between instrument and player conjures all the power and subtlety of a harmony that the bodily organism alone enables us to develop and to understand.

One may praise an artist for being, as it were, one with his instrument—in the suppleness of his arm and wrist, the touch of his fingers to the strings, indeed the grace of his posture itself gives one the impression that these too resound. And don't they, in fact? Who else inlaid the sounds within the instrument? Who else coaxed them out? Where else would they come from? Is it not the organism's own power and beauty that surges out at us, that seizes us by means of our selves?

Musical instruments are, after all—like the craftsman's tool, only to a higher degree—a continuation of the organ, a continuation of the whole human being. Could the wind instrument, held in the hand and pressed to the lips, form a closed chain of sound in any real sense with the whole human being if it behaved somehow contrary to the organ, if the body's limb had not imparted its own structure to it? Instrumental music that is "made" by the human being, even if less immediately than is true of singing, is also the organism itself resounding.

One does not "make" singing. It is therefore telling that the expression "to make music" does not mean that sound wells forth immediately from an organic harmony; nor does one say of a choir, as one would of an orchestra, that "the music swells."

Ever since the monochord cast a beam of light into the dark vesti-bule of the brain's nerve center, allowing the microscopic miracle of the organ of Corti to shimmer like a cerebral spectrum, an *introite* will go on resounding here, until the light of inductive research, aided by stringed instruments, eventually disperses the twilight that still shades the whole truth from us.

The history of stringed instruments since late antiquity shows steady progress in their manufacture, through the seventeenth century, when

188 The Fundamental Morphological Law

the violin reached the peak of its technical perfection. While its construction can never be governed by the golden ratio, it certainly puts the golden ratio to the test—which it passes unconditionally, in as surprising a manner as it does in the domain of the sculptural arts.

Fig. 42.

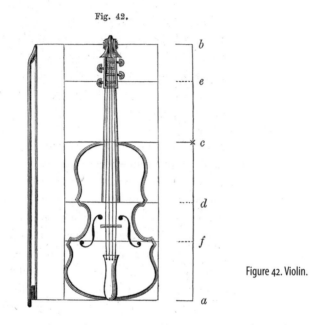

Figure 42. Violin.

If one were to sketch a surface view of a violin and immediately around it a rectangular enclosure, and if one then drew a horizontal line across the enclosure at the point where its length were equal to its vertical distance from the top of the enclosure, then one would find the sound box, neck, fretboard, bridge, and sound holes all at the correct distance ratio from one other (Figure 42). The same agreement with the golden ratio can be seen, without significant deviation, in the ratios existing between the soundboard, ribs, and back. The subtler, at this point almost unspeakable and difficult-to-define relations and differences among camber, sound post, bow, type of wood, and the set of the strings will need to be further clarified. This, though, will succeed only in tandem with the following insight: if the golden ratio's proportion is to be the measure by which we evaluate our conceptual world, beyond what is readily apparent to our sense experience, it will have to serve us in tracing ideal relations—if, that is, it means to be adequate to the unity of the sensuous and the spiritual that is realized in every authentic work of art. Recently, there has been a new method developed in the manufacture of violins by Professor Tuzzi, which has sparked some debate. The problem with manufacturing new violins—which is the same problem that the old master craftsmen confronted, with respect to euphony—is said to have been resolved by an as yet undisclosed process that is supposed to impart the proper qualities to the materials. It used to be that the wood from which the violin's soundboard was made required anywhere from sixty to one hundred years to complete the drying process, from the outside in. Now, by means of a new invention, the same process can be completed in a shorter time, and from the inside out, by pumping heated air into the wood's pores so that sap and bits of resin are extruded. Even drying among all the layers of the wood is supposed to produce uniform vibrations, without negative effect on the wood's vessels, cells, fibers, and knots.

We can see that this new invention—which has yet to stand the test of time, as have the soundboards of Amati and Stradivari—is not concerned with the external proportions of construction, but with the inner proportions of the material's molecular structure. Should molecular structure be somehow exempted from the universal fundamental ratio?

The old master craftsmen worked more instinctually, helping themselves to wood from old church pews and household furniture; today's master craftsmen profit from science that promises to impart to young wood the properties of the old. But it must be left to the future to determine how compatible this understanding may be with organ projection, from the standpoint of the fundamental organic ratio.

Seen this way, the violin soars far beyond the sphere of applied arts into that of fine art. And, incidentally, it ranks alongside works of art in terms of monetary value as well: only recently, a *Cremonese* violin by Stradivarius that belonged to a cathedral in Prague was sold for several thousand gulden. The tonal mystery the old Italian masters once intuitively breathed into their instruments remains aloof still to even the most meticulously crafted mechanical imitations.

But now there is a prospect that newly galvanized interest in Zeising's theory could help pave the way to the revelation of this mystery—an assumption enormously strengthened by the interest that Caspari invests in the subject. "In the golden ratio," he writes, "too much *similarity among the parts* (as too weak a contrast) and too much *dissimilarity* (as too high a contrast) are to be avoided. And so it seems reasonable to seek in this expression a subtle, spiritual, mathematical measure of value that may provide an objective point of reference for the different kinds and forms of intellectual activity, not only in terms of aesthetics but also in epistemological terms as well as in regard to psychology. Our explanations above have already shown how our measure of value has suggested a number of means for sublating one-sidedness, logical impossibilities, and contradictions with respect to the formation of concepts."[40]

Prior to Caspari, Conrad Hermann had been the only one to have vindicated the golden ratio in the field of logic, by putting the universal law of aesthetic harmony on a level with the logical law of correctness, then recognizing, by the same token, the logical form of inference in the golden ratio, and by proving that the three judgments comprising an inference correspond with the three members of the golden ratio.[41]

Already in his earlier work on *The Prehistory of Humankind*, Caspari's section on "The Emergence of the Art Idea" culminates in his realization that the golden ratio reflects the nature of aesthetic harmony and the basic aesthetic idea of the universe. Here too is expressed, for the first time, his unbiased conviction that this law is likewise binding for the philosophical idea of truth. Such identity with the philosophical idea stands the law on unshakable ground.[42]

The detailed analysis above of the most perfected musical instruments obviates the need to present further examples of the relationship between the applied arts and its neighboring spheres of production. Should the reader nevertheless regret their omission, there are specialized illustrated historical works on the topic, to which he is referred for further information. These will help to facilitate his understanding of individual products of artistic techniques, with respect to how the greater or lesser degree of perfection in their modifications correlates with the golden ratio. As Caspari remarks, "It is not as if one will be struck everywhere by the mathematically absolute expression of the golden ratio; but organic development and the law of formation seek to abide by it, in order to prevent forms from *deviating too extremely* from this fundamental proportion."[43]

If authentic art is found only where the laws of beauty, emerging from the human figure, are exhibited in the sculptural conjugation of an interior with an exterior, then it follows that the golden ratio is immanent in the artwork—that is, it is not something alien brought to it from without. That said, this is the verdict to which every attempt at establishing the rules of art must answer. For art cannot be subject to outside disciplinary measures; it is a measure unto itself and metes out its own rule to the boundless terrain composed of its after-images, a terrain in which artistic craftsmanship reigns.

At the same time, one must always keep in mind that, because great artists are often originally trained in the trades and the applied arts, art itself can at times experience a decline into a workmanlike state of adulteration. These transitions, which for the most part take place imperceptibly, nevertheless all involve a threshold relating to what we have called metrics. Where metric ceases, art begins, and where rational measurement begins, there art's incommensurability ends.

While measure inheres in the innumeration, in the organic on its own terms, metrics are mechanically appended from without.

This in no way means, however, that artistic splendor does not on occasion break triumphally through into handicraft and into the applied arts especially. Often we have seen how an inspired idea can lift the clever worker from his anonymity and make of him a great inventor, an acclaimed artist.

The instruments and equipment we have thus far discussed are objects that the human being can do without on occasion, without threat to his existence, for they are less immediately adjacent to his body. When it comes to bodily coverings and their perfection in the *costume,* the situation is a little bit different. The reasons why we also consider *architecture* in connection with costume should become readily apparent.

Because it is not feasible for us to go into great detail, let it suffice to say that the concepts of both costume and architecture evolve out of apparatuses originally designed to protect the body—*clothing* and *housing.* The expression "portable dwelling" for a body covering is more than a merely figurative indication of the basic quality they share in common.[44] The affinity between the terms "habit" and "habitation" is also indicative of this, as is the euphemistic phrase "wooden dressing gown," meaning the final resting place of kings and beggars both.

As far as clothing is concerned: the waist (*taille,* or *entaille* in French, literally the "insection"), which sets the tone for what is considered flattering in contemporary attire, aligns with the golden ratio's principal partition of the human figure.

In his book *Cosmetics, or the Art of Human Embellishment on the Basis of Rational Hygienics,* popular writer Hermann Klencke finds the golden ratio so pervasive, even in fashion, that he is compelled to remark: "What we have said should suffice, for popular literature at least, to convince the reader that the regular, agreeable, and handsome body shape is not

Fig. 43.

Figure 43. Female costume.

exempt from laws pertaining to *spatial part-whole relationships* that are in force elsewhere in nature, laws whose formal expression is pleasing to the refined and cultivated sense."[45]

Wittstein offers a much more in-depth analysis of clothing, including illustrations, in his own related work from which Figures 43 and 44 have been taken.

In the context of the performing arts, the concept of "costume" extends beyond the forms of clothing that cover the body to include everything the body carries, the hand-crafted furnishings that populate

Fig. 44.

Figure 44. Male costume.

the living space, and even further out to the adjacent environment. Therefore workshops, studies, bath rooms, parks, royal courts, auditoria, libraries, theaters, and so on are part of the costume of any individual who
occupies these spaces and is thus characterized through them.

In 1876, *Der Salon* published an informative review of August von Eye's
Applied Arts of the Modern Era, displaying deep insight into commercial
circumstances of today. Von Eye's understanding of the relation between
living space and attire is entirely consistent with the above: "We are mindful above all of what surrounds us, our environment and our apparel, for
herein we are at home, herein we suffer and rejoice. . . . Wherever we
feel at home, we will make an effort to arrange ourselves such that each
expression of our being, even those not immediately adjacent to us, will
put us at ease—thus, living and bed rooms, house and garden form a
continuation, an *extension of our clothing*."[46]

It is worth remarking, with respect to architecture, that Zeising tested
his theory against ancient and medieval structures, temples and churches
in particular, and came up with a considerable amount of evidence in
support of the accuracy of the law of proportion. Detailed illustrations
accompany his argument here as well.

As the concept of costume evolved, as it came to signify something
beyond the body's original need of clothing and shelter, architecture too
passed beyond the provision of ease and utility to become an act of free,
monumental configuration through which the human being becomes
conscious, as he does through all works of art, of the organic idea incarnate in him, yearning to manifest.

It follows from this that all of culture is in actual fact the grand costume of humankind, under the all-transfiguring authority of the law of
proportion that radiates from the human figure and is alive in the golden
ratio. We agree with Zeising's statement that *culture is nothing but self-
perfecting human nature in a higher sense*.

Let us now review what we have thus far discussed concerning the
fundamental organic ratio.

In contrast with mechanical construction, organic organization involves
the reciprocal accommodation of all members to one another, such that
each member guarantees the existence of all others, in keeping with the
theory of the golden ratio.

Recent literature concerned with an evaluation of symmetry—which finds its highest expression in the human figure—has come to recognize the significance of this theory. In whatever context these works assume the human being as the measure of things, they relocate the concept of measure into the living unity of organic distinctions, into the self-reiterating standard proportion, meanwhile relegating metric and number to the sphere of mechanical forms, where they properly belong.

All artifacts are manufacts, works of the hand. Handicraft, the applied arts, and fine art represent stages of increasing perfection in the transition of the artistic from the artifactual to the work of art, from the useful and pleasing to the purely pleasurable, from instruments to free creations that exceed all necessity.

In the hand, the law of proportion has the greatest latitude because it is the working organ, the one that is most flexible and most capable of giving form. The hand represents in its actual accomplishments—and even symbolically, as the French word *manoeuvre* suggests—the whole human being. Mind and theory are on one side what hand and praxis are on the other. The goal of all contemplative work is that brain and hand, theory and praxis coincide all around. The human being nears this goal by equipping his hand with the means adequate to expand its power and dexterity beyond bodily limits.

The hand accomplishes what is to its benefit and always remains true to itself in the adaptation and handiness of its constructs, while the proportion in its articulated conjunction becomes the measure for the artifacts proceeding from it. There is a saying: to do something "off the cuff," or "offhand"; philosophy and mathematics could hardly dream of the things the hand does in this way.

Against the hidden or disavowed backdrop of an acquired empirical knowledge of the phenomenal world, the imagination inflates, forced into so-called pure abstract concepts that then of course immediately burst. It is the same with number as it is with space and time. In empty infinity, ultimately everything escapes.

But the concept of infinity is different where there is a prospect of appearance—that is, the idea of the infinite, or the infinite thought as the eternal becoming of the finite.

The definite article—when, for instance, we say "the" space—already reminds us of the backdrop and colors the so-called absolute void. "The" number also appears in a particular context. Presented schematically as line, in one direction it remains the unity of distinctions, as the conjugation of kind (the *artion*); in the other direction it is always a sequence of discrete, supernumerary units (the *perisson*).

On the one side, then, *innumeration*: organic organization and distinction, art and philosophy. On the other, *enumeration*: division and numeral, mechanics and mathematics. And midway between interiorization and exteriorization, hand in hand with speculation and exact method: the *formula,* the great question posed to the objective world—which not only answers but is in fact the very source of the question, and this is so because the formula coincides with the world's own mute question posed to the human being, the one who *thinks* and *measures*.

This review should suffice to fully legitimize the natural tool, the hand, as an organ in the superlative sense. Before turning to another aspect of our study, however, it is in our best interest to examine the circumstances surrounding the hand's mystical encounter with the pentagram, in order to ensure that the hand not only gets clear of the mist but that it also acquires some prestige from the encounter. For we will have frequent occasion to call upon the hand as our most credible representative of the standard proportion and as the guarantor of our counting values.

The human hand and the pentagon are similar enough in their proportions to have sparked lively debate in the past. But the matter seems to be a simple one, harboring hardly any more mystery than a blade of grass.

Were one to extend each of the five sides of a pentagon in a line in both directions until the point at which each of these lines intersects another extended line, then a slender triangle will be formed on each side of the five sides of the original figure, attached to it like limbs, such that when the whole is considered schematically, it may be compared, as it has been before, to a splayed hand. If one were then to draw all of the diagonals inside a regular pentagon, these would intersect one another according to the golden ratio's proportion, forming in their center a new regular pentagon, only proportionally smaller. Each side of the smaller pentagon is to the line extending it in a single direction as each side of the larger pentagon is to the line extending it in a single direction—in other words, each side of the pentagon is to its extending line as the minor is to the major segment of a line partitioned according to the golden ratio. Conversely, by drawing straight lines connecting the points of the triangles formed by extending the sides of the original figure, a larger pentagon is formed, and the extended sides of the original figure now form the finished diagonals of the outer figure, each of which is intersected in the standard proportion.

This procedure may be reiterated inside or outside any regular pentagon, with the same result—the continuous formation of mutually implicated diagonals and pentagons in the golden ratio's proportion.

The golden ratio appears, then, as the determining force in this reciprocally conditioned transition of one element into another. The five-count of the fingers and the decadic system are in this way absolutely indispensable to the ratio.

Conrad Hermann attaches such high value to decadology because the number ten, as such, is supposed to be of paramount and decisive significance for the inner structure of the reality that surrounds us. It has quite the opposite effect of diminishing our respect for the golden ratio, when we read in the closing remarks of Hermann's essay on the law of aesthetic harmony: "Ultimately, the arithmetical-logical and aesthetic-organic question turn out to be one and the same. And we believe we are beholding, in the number ten and in the rule of the golden ratio, the two key indicators for a complete reworking of the question as to a universal organizing principle for all of reality."[47]

The simplified symmetrical design of the pentagon shown here is nevertheless an aid to understanding the symmetry and the myriad distinctions in the prototypal organ (Figure 45).

Other than the air of the mystical which we discussed earlier, there is still another interpretation that has opened the floodgates to superstition and even occasioned some local public disputes. We are talking about *chiromancy*. This is on a level with *astrology* and *alchemy*. Just as the latter secretive speculations, admixed with empirical fragments, ultimately emerged from the darkness to become imposing sciences of the first rank, so too this display of the hand's lineaments, chiromancy, turns out only lately, and in agreement with our view, to be the "theory of the proportions of the human body." This is yet further evidence that the

Fig. 45.

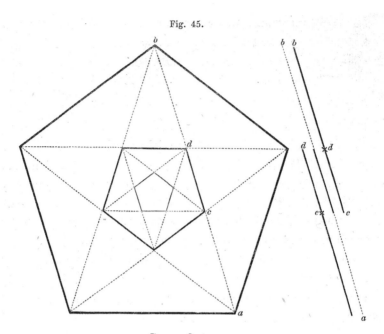

Figure 45. Pentagram.

sciences do not always proceed in rectilinear fashion but along looping paths, led by an intuition—which first must slowly extricate itself from the bonds of superstition before it can flourish into maturity—that there do indeed exist fundamental organic laws.

Chiromancy has never been quite as renowned as its sister sciences—astrology being concerned with the influence of the cosmos in human destiny, and alchemy with the telluric influence. But chiromancy relocates fate immediately within the human being himself, in his very own hand, in this symbol that accounts for his conduct [*Handeln*]. All along, the benefit in it has been its dark suspicion that the fate of the human being grows together with him "life and limb." The fundamental organic law, as fate's constituting power, can no longer be denied.

The words of the poet—"In your breast lie the stars of your fate"—are open to a new interpretation.[48]

It may be unnecessary to suggest that the number five is not only represented in the body's structure moving outward from the torso but is also reiterated uniformly in each individual extremity.

What the five fingers are to the arm and hand, and what the five toes are to the foot and leg, the five senses are to the head and neck. The localized sense organs—vision, hearing, smell, taste—function through involvement of the skin, as the uniquely sensitive organ of touch. The distinction that concerns us here has to do with the fact that, though each is individually mobile and their movement occurs in organized isolation, the sense organs nevertheless appear in nearly perfect reticulation.

The facts are: there is a connection among sensory activity, sensation, and natural speech; the tongue has a dual function of tasting and speaking; and there are also local bonds between the sense organs and the organs of speech. In keeping with these is the highly suggestive fact that sensory activity triggers speech and thought and, in turn, speech and thought inform and train the senses.

The tongue, the *lingua,* is what connects the realm of sensibility to the spiritual world of language, and the consequential innumeration of the five fingers and the five senses is associated with that of the five vowels, the original representatives of spoken language.

CHAPTER 12

Language

A s noted earlier, the depiction of organ projection advances from artifacts that are but crudely sensuous after-images of natural tools to those in which tangible material gradually gives way to an increasing spiritual transparency.

Given the historical character of the many different means of culture, often emerging simultaneously and somewhat confusedly, the discursive depiction is hardly able to proceed stride for stride, preferring instead to keep to relatively more complex and finished products. For this reason, it is only now that we may begin our discussion of language. We turn our attention first to writing, which, as *hand*writing, provides us a natural transition from the preceding section.

If we take script and scripture in the broadest sense to mean everything that records the spoken word for the eye to see, then printed matter is the immediate corollary of the written word. The one passes over into the other; the written word will be printed, and what is printed can again be transcribed by hand. The art of printing is nothing but a replication of the sign-making hand and of its dexterity, originally exercised with and on the plainest materials. What the high-speed printing press can accomplish is essentially the same as what the scribe once did, albeit the latter required greater effort and longer hours of work. Greater uniformity and time savings are the only outward differences.

The hand equips itself with materials for printing, just as it once equipped itself with materials for writing. The hand prescribes what the typesetter is meant to transcribe for transmission through the high-speed replicating machine of the press. Thus ancient Egyptian monuments and Mesopotamian brick, no less than the skillfully crafted technical products of the modern era, comprise the whole of human scripture. What we say of the telegraph—that *it writes*—the same can be said of the press.

All *handwriting, litera* and letter, is the manufactive portrait of spoken language. But the greatest of all manufacts is script. It too is the unity of a real and an ideal, a work of art whose magic encompasses and records the whole human being in the signature *manu propria,* the most succinct visible abbreviation of a personality.

In language, the otherwise firmly established distinction between artwork and tool falls away entirely. Because language explains exactly what it is, it is in effect practicing exactly what it wishes to explain. Thus language is the tool through which it understands itself as its own tool—a spiritualized tool, both apex and agent of the human being's absolute self-production. A thought-form, in the sense that form itself is thought and thought form, language is also the unity of an ultimate distinction. Depending on whether it appeals to the ear or the eye, we distinguish between *spoken* and *written* language.

Comparative linguistics has traced the development of its object all the way back to a primary consonance between gesture and sound, discovering the *roots of language* in "sound gestures" and potentially proving that speaking commences with the organism. Just as hearing, seeing, breathing are each predisposed to a corresponding organ, so too are sound gestures an organic function; they are in a sense the echo of the organism. Therefore, on the basis of the fact referred to earlier, that *the organ's functions are a manner of appearance of the organ itself,* August Schleicher holds that language is a real existence, a natural being. What some have called the linguistic drive, the power of language, the need for language, and what Jacob Grimm refers to as the "unconsciously governing spirit of language," is innate to the organism.

Understood this way, one is less inclined to take exception to a certain prevalent, rather hardline expression: "Man is language," or man is—instead of the usual "has"—consciousness, reason, and the like. It is therefore assumed that whatever the human being *is* from the outset, as an indivisible complex of organic activities, this he cannot *have* nor have appropriated to him as some sort of extraneous possession.

It makes no difference whether the emergence of language is attributed to an unconsciously governing spirit or whether, in the words of Adolf Bastian, "we do not think but it thinks in us." For this "It" is always the unconscious, simultaneously operative in both thinking and speaking; it thinks in us and speaks out from us.

The *principle of alternating effects,* which Ludwig Noiré rightly emphasizes, explains that thinking and speaking are one and the same.[1] The assumption that one has priority over the other is groundless. There can be no such one-sidedness—neither at that first stage in which language forms are, in Eduard von Hartmann's words, "the collective effect of a mass instinct," nor later, once language has become relatively fixed through writing as something we deliberately acquire and that is conducive to international exchange.

The simplest expression for the unity we have identified is the "It"— a thing that continuously strives, in the transition from sensing to representation, to become the I that thinks language or that languages thought. For, behind the It we find the human being's entire spiritual nature, the psyche, the I, the personality.

If, in *The Organon of Cognition of Nature and Spirit,* Carl Gustav Carus refers to speech sounds as the equivalent of spirit, then the complete word—whereby the representation, in binding a thing, is reified itself— must be that much more: the immediate unity of the sensual and the spiritual; a unity of sound, of sense, of outer and inner speech-forms; a unity, in short, of organic distinctions, just as we repeatedly saw it confirmed as innumeration above.[2]

As long as the "I think" and its procedure fail to be recognized as the functional relation of body and spirit, this indeterminacy of the functioning I will be provisionally displaced into the It, in which are hidden the absolute, the unconscious, the will, primordial consciousness, and whatever other name we may use to refer to the divine. The It is the unfathomable universal Something of metaphysics—a Something that escapes the shackles of every dogmatism, a Something the human being may approach only insofar as he minimizes the sphere of what is unconscious to him while at the same time and to the same extent expanding the sphere of his consciousness by advancing in knowledge of himself. This is the process of alternating effects: consciousness is positively reinforced through immersion in the unconscious during states of sleep and rest, on the one hand; then, on the other, this rest renews the energy the human being requires to carry on further unlocking the unconscious.

Thinking and self-consciousness extend only so far as language extends, and vice versa. It is on this basis that we have to investigate how language, as a product of organ projection, acquires an aspect enabling it to occupy its rightfully privileged position in the category of "tool." In order to do so, we will have to attend both to the material in which the bodily organ is unconsciously re-created as well as to the fact that comparison of the after-image with its prototypal image is what establishes it as a tool, by means of which we improve our knowledge of the bodily organism.

Everyone is more or less familiar with the vocal and speech organs. And Schleicher's definition of speech—that it is the aurally perceptible symptom of the activity taking place in a complex of material relationships formed as the brain and speech organs develop in connection with the affiliated nerves, bones, muscles, and so on—is clear enough. Still, there are many who search in vain for the substance in which speech is meant to take shape. This substance, though, lies closer than one might suspect, even if it is less concrete than what we typically think of when referring to "substance." In the conclusion to his lectures in physiology, Johann Czermak explains as follows:

"Song and speech are composed of elements that, considered objectively, are actually nothing other than acoustic phenomena, propagated from the mouth of the singer or speaker to the listener's ear, through the space between them—without there being anything inherently having to do with spirit involved.

"What fills the space between mouth and ear—though invisible, not unrecognizable—is a meaningless, *purely* mechanical *surge of sound waves!* It is only in the brain of the listener capable of comprehending that the transubstantiation takes place, of the absolutely material movements of sound, whether voice or speech, into the physical state of sensation, feeling, and thought.

"Thought and feeling—vocalized—become in actual fact *matter in motion,* and this is transfigured in turn, in the consciousness of the listener, into feeling and thought.

"The path from one consciousness to another traverses the despised coarse material world, directly and—it is so even for beautiful souls— without mercy."[3]

The substance in which speech sounds take shape, then, is the air. The air's movements, effected through human breath, are mute and invisible sound waves that, upon encountering the ear, are transformed into the sensation of tone. The sound wave's vibration ratio, depending on amplitude and duration, determines the intensity and pitch of the tone.

This said, we can see the path ahead along which spoken language becomes a tool, and it lacks not a single element necessary to the concept of organ projection—neither the prototypal image recognizable in the functioning organ, nor the mechanical after-image in the air's vibration, nor even the unconscious process of its formation. All that remains for us is to prove that the unconsciously emerging after-image, once consciously perceived and compared with the organic prototypal image, may be used to enhance the human being's knowledge of physical phenomena.

We are so accustomed to using the concept "tool" in a figurative sense that its real and serious application is seldom properly appreciated.

Therefore when, among others, Wilhelm Wundt describes language as a "convenient instrument" and as the thinker's "most important tool," and when William Dwight Whitney writes that human beings as a species invent language, the organ of their spiritual activity, an organ that serves them just as well as the mechanical apparatuses by means of which they assuage the burden of and intensify their capacity for physical labor, they both mean "tool" in the literal sense to which we are strictly adhering.[4]

There is, however, a distinction to be made between the spoken and the written word, as tools. The spoken word does not appear as obvious or tangible as the written word does; rather, in keeping with the invisible and impermanent material that Czermak nevertheless refers to as coarse, the spoken word appears in the form of sound wave vibration ratios. These sound waves, generated by emanations from the speech organs, have to legitimize themselves as sound and thought before the jury of the senses.

The distinction thus boils down to the following: it is the ear, rather than the eye, that functionally, sensibly interprets the structure of the sound wave; as a result, it is the ear that is bound in solidarity with the vocal organ. The information available to us by way of physics—concerning periodic sound waves, how these are measured and counted, and the diverse ways in which, in an instrument, these combine to produce basic

tones as well as secondary and overtones—is highly useful for our understanding of the physiology of the speech organs.

The real relevance of Helmholtz's discovery of the composite nature
of sounds, when applied to the physiology of speech sounds, is that *vowel
formation* is also based on the consonance of overtones—vowels being,
in Czermak's words, "nothing but the different timbres of the human
voice that emerge when the oral cavity assumes different shapes to intensify, via resonance, certain overtones in the sound of the voice and to
dampen others."[5]

According to the difference in the sound—the sound being either tone
or noise, depending on whether the air's vibrations are regular and periodic or irregular—speech sounds may also be distinguished as vowels
and as consonants, or, in other words, as tone forms and as noise forms.

In the third of his *Lectures on the Science of Speech*, in which he deals
with the physiological alphabet, Max Müller explains the development
of "vocal tones" and "consonantism." Here he provides an account of
what forms sounds (vocalized or non-vocalized breath), how they are
formed (by opening or narrowing the glottal ligaments), and where they
are formed—namely, "by active and passive organs in various positions,
the standard positions being those indicated by contact between the root
of the tongue and the palate, the tip of the tongue and the teeth, and the
upper and lower lip, in addition to a number of possible modifications."[6]

It is apparent from the above that one cannot discuss the organs of
speech without reference to our earlier discussion of *wind and stringed
instruments*. For the latter are projections of the organs of vocalization
and hearing; in other words, they are the apparatuses through which the
general aggregate of organic sounds, immanent in human enunciation,
achieves a mechanically specified expression. Even the talking automaton, this ridiculous figure of wires and gears, is basically just a perversely
malformed musical instrument.

It was this preliminary stage of inquiry into the instrument that informed us of the substance at work in the formation of speech sounds.
This substance, its formation being ephemeral, also provides a tool
enabling our further inference into the process of thought and into the
truth in Hegel's expression that "it is in names that we think."

The stringed instrument first informed us how the ear is physiologically appointed; the structure of the musical organ first provided insight

into how the lungs, trachea, and larynx cooperate in the process of vocalization. But the arrangement of strings and the organ of Corti, the organ's
pipes and the trachea, are, like all mechanical and organic structures,
distinct from one another. In speech, on the contrary, tool and operation are spatially and temporally simultaneous, and this is why word is
thought and thought is word. "In music," writes Helmholtz, "it really is
the sensations of tone that form the substance of art"—and in language?[7]
Wouldn't those elementary sounds that combine to form words actually be the sensations of things that form the substance of language? We
may refer the question to Wundt's remark that, with speech sounds, each
steady timbre, whether of tone or noise, has become an element within
a diverse array of representational and emotional signs.

The representation that the human being has of himself in this way,
in speech sounds, is quite correctly described as self-consciousness.

In a recent issue of *Gartenlaube*, Ott. B. shared an extremely interesting linguistic fact from his own experience, a fact that promises to sub-
stantively enhance our knowledge of the formation of speech sounds.
This has to do with the pronunciation of the English language, with
which foreigners have such difficulty. "But," he assures us, "it is not hard
to perfectly imitate English pronunciation; one need only remember to
jut forward his lower jaw. The desired shift in sound will occur immediately. As if by magic, the German consonant *w* will be transformed into
the English vowel double-*u*; our hard *th* becomes the ineffably soft British lisping sound; even our rasping *r*, which English does not have, disappears on its own."[8]

With jutted chin, the tongue is forced forward to play along the back
of the upper row of teeth; it therefore has a much shorter distance to
travel. Rather than the brusque *all, alk, aw*, the English half-diphthong
aoh is sounded, the hard *k* of the German *Knie* is dropped to become the
English *knee*, and so on. Among the English, who for the most part have
prominent noses and chins, we see that a long, narrow nose necessitates
a long, narrow tongue, and this in turn a lisped toothed speech. Among
the Celts, Irish, and French, short, broad noses and tongues are more
common, resulting in more of a palated speech. The chin plays the principal role in speaking and, as a rule, the tongue is proportional with the
lower part of the nose. A rapport exists between nose and tongue, not only
where the sensations of taste and smell are concerned—both of which

are served by the same nerve of the fourth cranial vertebra—but also with respect to their dimensions. With such a brilliant physiological foundation provided for the problem, the article's concluding suggestion would seem to be perfectly justified: there is something to be developed from this "linguistic egg of Columbus" that may aid in explaining Grimm's law of the First Germanic Sound Shift.

We are convinced that sound formation is governed by the natural spatial relationships among the organs of speech and by the kinetic no less than the kinematic links among them. Moreover, we are convinced that the Sound Shift in particular is the result of a kind of shift, for the most part inexpressible and occasioned by habit, in the elements comprising these links. To understand this shift, we have to take ancestry, soil, climate, habit, way of life, and nutritional characteristics into account, as having a role in determining the shape of the teeth, tongue, and mouth.

Self-consciousness could not have developed if head and hand had not continually moved to accommodate one another—that is to say, if technology had not supplied stimuli and substance to the need for language, and if language had not imparted to the need to give form conscious awareness of its capacities and how to perfect them.

Our investigation to this point, tracing the course of the history of civilization in broad strokes, has led to the argument that organ projection, which takes form in the world of artifacts created by the human being, determines the unfolding of self-consciousness.

Now, if the same process declares itself in the development of language, this is essentially distinct from all other forms of organ projection in that language is free of the inhibitions that otherwise, when dealing with solid matter, manifestly resist the need to give form. Moving in the most pliable and, in a sense, spiritual element, language allows both the tool and the organic activity that shapes it to appear as uniformly cooperative. One can therefore say, without paradox, that thinking is just as much the tool of language as language is the tool of thought.

Accordingly, if we wish to establish the relation of language to all other contents of culture, then we might say—borrowing Wundt's expression "that the soul is the inner being of the same unity that we perceive outwardly in the body belonging to it"—that language is the inner being of the same unity that we perceive outwardly in the world of culture belonging to it.[9]

We now turn from this highest conception of the tool, appearing as language, back to the element that links the spiritual with the coarse material structures of human self-production, to *writing*, in order to demonstrate that it is only when speech sounds are present to memory in visible signs that the requirements of the need for language are met in full.

As we did when the electrical telegraph was the object of our reflection above, here we are initially dealing not with the nature of the written character but with the machinal design of the writing apparatus. Although the motive force of the dictating organism and that of the mechanical activity conveying the dictates is ostensibly the same, we must nevertheless be vigilant about not confusing the organic performance with the mechanical apparatus. For inorganic matter, even if originally the same as that which comprises the mechanism, is fundamentally changed once it has been incorporated into an organic existence and for as long as it continues to participate in the acts of a living will. The telegraph, a lifeless piece of machinery composed of inorganic matter, is transformed through an act of human will into the bearer of thought and language. The electrical current in the body of an organism is fundamentally different from the electricity occurring sporadically in the different forms of individual mechanisms. The same is true of the tonal world contained in the human voice, whose different sounds appear mechanically reproduced in particular kinds of musical instruments. It is by means of the conscious will of the human being, not any sort of electrical cloud, that the telegram materializes. The telegraph therefore clearly belongs to the history of writing materials and to *The History of Script and Writing*.

This is the title of the first volume of a work by Heinrich Wuttke that, as writing about writing from its "crude beginnings in the art of tattooing to the laying of electromagnetic cable," is proof of its own claims, from the standpoint of a genetic survey of the object.[10]

Since any discussion of writing has to deal with the materials of character representation in addition to the characters and their meanings, we will remark only briefly that the history of the electrical telegraph must necessarily be traced all the way back to the first stylus-like object ever used by a human being. The organic prototypal image of the first stylus, then as now, is the finger, which, in absence of all else, has had to do the work of indicating and explaining. This archetypal image of all inscribing, sketching, painting, and writing tools remains recognizable in each of its manifold forms.

It does not matter at first what material is used to record the sign, nor what material the sign is recorded on—whether bark, stone, metal, wax, velum, or paper; nor does it matter whether the sign is made visible by applying substance to a surface or by cutting substance away from one. The crucial thing is always the image of the fingers acting, in their natural propensity to write, as stylus, pen, reed, needle, chisel, quill. The pressure of the telegrapher's fingers and that of the fingers guiding the quill are not essentially different. Even the roll of perforated telegraph paper is a manuscript that may still be converted back to an original form of script through transcription with a quill before then being sent off by dispatch.

If spoken language extends into handwriting, then telegraphy and the printing press are handwriting's natural extensions. The corollary of the speech sound is the letter, in German, *Buchstabe*.

Lazarus Geiger's etymological commentary on this remarkable German word correctly relates its concept with speech sound while at the same time presenting a striking example of "how flimsy a merely phonetic etymology can be if it is not bound to conceptual laws."[11]

According to Geiger, in German antiquity "buch" meant a written thing, irrespective of the material on which it was written. "*Boca* (*boec* in Anglo-Saxon) first meant letter (graphic character), then script, then epistle, while *bōkāreis* meant scribe. One may be tempted to interpret the second part, *stab*, as rod or staff, and thus *Buchstabe* as a booking or accounting staff; but the primary meaning of *stab* is the same as that of sound. We know this for certain through comparison with the word for voice (*Stimme* in German, *staefn* in Anglo-Saxon, *stibna* in Gothic), whereby the shift in meaning follows the same pattern as *vox*—cry, voice, speech sound, word. It should therefore be clear that *boec* and *staef* are essentially tantamount and therefore interchangeable with one another. The former means letter, *litera*, the latter speech sound, *vox*; both together fairly correspond with the Latin *literarum elemental*."

For our purposes, though, we will hold precisely to the distinction between *litera* and *vox*, between the written sign and the utterance, the graphic character and the speech sound. Combined in a single concept, the words actually mean *written sound*—that is, the sound as it is appears when written, or *Buchstabe*.

The etymology of this word, then, has given us the basic idea behind both the emergence and alphabetic development of script in general as well as the development of the specifically Semitic alphabetic script from pictographic writing. The German word *Buchstabe* means the unity of speech and writing. In the complete concept of language, text and sound are inextricably interwoven. We are therefore tempted to assume that the elementary signs of written language somehow coincide with the elementary sounds of spoken language and that, moreover, as with sound gestures, the sign-making of writing's incipient phase could be readily described as "graphic gestures."

The original coincidence of gestures with natural sounds, as an expression of sensation and of nascent representation, may well have corresponded with a similar association between gestures and fixed external signs—like, for instance, the hand raised in warning and the tree branch set up to signify the same, like the gestures of mourning that extend, in written form, into our memorials to the dead.

In script, speech is made permanent. Script is sound enchanted, the unity of sound and sign that the letter literally means. At any moment, the sound may echo again out of the letter and, as a living word, captivate the spirit of the listener. In short, the letter is the symbol of an indestructible solidarity, of the reciprocal immanence of what is heard and what is seen, of the graphic character *(buch)* and the sound of the voice *(stab)*, of speech and script—in a word, the letter contains language in its entirety!

In this indivisibility, each work of language [*Sprachwerk*] appears as a product of the functions of the brain, the hand, the tongue—or whichever member in the concatenated set of speech organs one might select in place of the tongue. It is evidence in itself that, on the one hand, script is a fluid factor in the ethnographic classification of humankind according to the differences in their alphabetical forms, and that, on the other, the graphic character, as an instinctive creation, is exactly that—a *character,* not a cipher made up and agreed on arbitrarily and substitutable at random. The Greek word for "character" originally meant the figure or text etched in or imprinted on stone, metal, or wood, signifying a unique (inner) distinguishing feature. Letters are therefore characters, phonetic signs with which the configuration of clan organization is somehow very closely related.

In keeping with the presumption that, in all cultural environments, alphabetic script emerged from pictographic writing and that the image became everywhere the sign and signification of the letter, it is necessary to infer a universal correlation—a "congeniality"—between the system of writing and the respective speech-organism and race-organism. The worlds of speech will differ, depending on whether the language spoken is monosyllabic, multisyllabic, or inflectional; corresponding differences will appear in their writing systems. Writing, however, does not have the same effect on *lingua* as either sound or the organism as a whole does.

If, therefore, speech lives in the human being and develops in him as a natural power, then his handwriting is the signature of his lineage—that is, the natural determinacy that marks him, in the universal and the particular, according to race and nationality and of course individual temperament. The proverbial "Speak, and I'll tell you which spirit's child you are" now stands coequal with the information we are able to glean about the character of a person through the character of his handwriting.

More than mere hobby and curiosity, the practice of collecting signatures is of real interest to physiology. Even the original manuscripts of great works of literature turn out to be world-historical national signatures and objects of ethnopsychology!

We have traced the process of organ projection as far as it extends to discrete mechanisms. All that remains to us, then, is to regard the last and greatest mechanical structure, as it loses itself in the spiritual, in order to see how it becomes the tool of insight into organic functions as well as the tool of the human being's self-conception to the highest power.

Because the way in which we understand language as a tool appears more sensibly perceptible and permanent in handwriting than it does in speech, the literature of a people is thus the quintessence of its works of language, in written form.

With only a cursory glance at script as handwriting and at the hand itself—whose organic activity produces signs in which meaning and image are intimately intertwined—it can hardly escape us that the fundamental law of the golden ratio, alive in the hand's organization, finds expression not only in the familiar graphic characters of contemporary civilized peoples but also in all typographical equipment. Wittstein touches on this topic in greater detail, sparing us the need to do so.

Unlike handwriting, which is an instinctive creation, its surrogates, stenography and code, are contrived through deliberate selection and always liable to random substitution. Therefore these surrogates must be evaluated exclusively in the following context: as entirely arbitrary signs for the alphabetical symbols emerging from and integrated with spoken language. These are the grounds on which every invented universal written language must be evaluated.

While attempts at inventing a universal spoken language have always failed in the past, and will continue to fail into the foreseeable future, experience with stenography, telegraphy, and code may be giving us reason to hope that a solution to the problem of a universal written language could be on the horizon, at least according to extensive preliminary studies undertaken by Dr. Phil. J. Damm in Stockholm. In the interests of global commerce, we wish the endeavor the best of success, for such a language promises to save a significant amount of both time and money.

Apropos of enciphering, we should not fail to mention Joseph Faber's "fabulous talking machine." A skillfully executed artificial reproduction of the anatomy of the human speech apparatus, it is a vivid representation of scientific findings and, as an unexpectedly amusing educational aid, it far surpasses all the so-called automatic curiosities of times gone by. All those old chess-playing, music-making, babbling, unedifying frames [Gestelle] may be left to their ordinary fate—the curiosity cabinet and the junk room! Though it's true they have not been completely useless. If nothing else, these mechanical gadgets—contrived to excite our momentary astonishment at the exploits of human wit, though far from arousing our enduring admiration—do provide insight into just how great is the gulf separating them from the organic tools that proceed unconsciously from the human being's need to give form and that therefore serve his cultural aims. Beside "gods' gifts" of spoken and written language, both shorthand and chattering androids can only be caricatures.

Linguistics and the natural sciences have succeeded in deducing the secret of speech sound formation. Very different types of strictly localized muscle movements, from the lungs to the lips, project themselves through the air in just as many different patterns of vibration. Once these are interpreted as speech sounds by the ear, they are then referred back to the activity of the muscle that projected them, resulting in an understanding

of the correlation between the sound itself and the local distribution of
these muscle movements.

Sound wave vibration ratios, originally discovered in the course of
investigations into the nature of musical sounds, have become the scientific basis for research into vowel formation and the formation of speech
sounds generally.

The relevant findings have proved wonderfully reliable when applied
to deaf-mute education. For the zones of speech sound enunciation are
specified so precisely in the mouth and throat that the deaf-mute stu-
dent can read the outwardly perceptible muscle movements of a speaker
as if they were written characters.

The papers recently reported the following truly astonishing success
story: *"At the Philosophical Faculty in Heidelberg*, a candidate sat for the doc-
toral exam under circumstances that may perhaps be called unique. A
few weeks ago (July 1875), a young man arrived, visited several professors, and made known his intention to sit for the academic examination in the natural sciences. This sort of thing happens so frequently that
nothing about it would have been remarkable had the candidate not carefully inquired as to whether any of the professors acting as examiners
had . . . beards! Was he betting on greater indulgence from beardless
instructors, on account of their similarity to women? Soon enough, the
real reason was revealed: the young man had been born completely deaf
and learned to speak not by listening but by another method, and this
practice he had developed of reading what was said from the lips of the
speaker made it so that, from the cadence of his speech alone, hardly anyone ever noticed he was deaf. A person with this sort of energy and dili-
gence certainly has the right to pursue a scientific career, and it may be
freely assumed that he passed his exams with the highest possible marks."

A comment by Professor Conrad Kilian on deaf-mute education should
suffice to explain as well as to remove any doubt as to the possibility of
such an event: "Fixed resonances in the oral cavity physically enable not
only a musical cadence in the deaf-mute's spoken language but also
expand its range by several octaves, causing harsh dissonances as well
as an irritating monotony to disappear. Vowels emerge in accordance
with the law of cavities discovered by Franciscus Donders, which states
that the pitch of the vowel sound, whether high or low, corresponds with
the width of the mouth's cavity, whether wide or narrow. Each vowel has

its own unique resonance, its so-called mouth space, which lends it its typical range and the lushness of its timbre."[12]

One would be right to designate this method of deaf-mute education, as it has evolved in Germany in connection with the study of phonetics, the organic method, whereas institutions outside of Germany prefer to use an artificial sign language and finger spelling that, while preserving communication through natural gestures, has more to do with the mechanical. The above-mentioned correlation between spoken language and script is again thrown into sharp relief here. The air's vibrations, heard as sound by the listener, the deaf-mute sees in the form of muscle movements effecting these vibrations. This seeing of sound occurs almost as quickly as sound is heard. Since each speech sound corresponds with a particular muscular articulation and since, in the course of a minute, about six hundred sounds may be spoken, this seeing of sound is a skill that permits one to read an equal number of muscular contractions at the same time.

Finger spelling, as a surrogate for speech, is a language of gestures that have passed through phonetic symbols. It acquires an element of visibility from writing, an element of ephemerality from spoken language, and it also takes something from the universal gestural language, in that it is gesture reduced to a single specialized organ. Behind the distorted forms of authentic writing, including code and telegraphy, stands the power of the letter; and behind the letter, the all-powerful speech sound, the veiled *logos*. From the *logos* all forms of language proceed, no matter how different they may be, in their external or internal form and the content of their thought. And to the *logos* all forms of language return, as both a consummate means of instruction and the rich subject matter of learning.

Now, if we were to disregard all internal and external distinctions and take up language as a whole, as a uniquely human property, on the one hand it would appear as a tool, to be used for the purposes of instruction and comprehension, and on the other it would appear as a product, by virtue of the universality of the content it communicates. With this endless revolving and passing over of the tool into the product and the product into the tool, the distinction gradually begins to disappear, as knowledge advances along ever more complex stages of self-consciousness, until finally the distinction vanishes.

All throughout the domain we have been traveling, the tool, as it emerged, could be more clearly distinguished from the objects on which it acted the more sensuous was the material of which both tool and object were made. As the material became ever finer, to the point of dissolving into breath, the distinction faded into darkness, and out of this darkness a pure manufact once more emerged beaming, in handwriting—only then to withdraw entirely, at the end of our journey which now we are nearing, into the general sphere of language.

Above all, in language as speech (*reda,* Old High German for "accounting"), in the deliberate expression of thought, the human being gives an account of himself as I. The I, the personality, is language to the highest power—that is, internal dialogue. In internal dialogue, tool and work, means and end, subject and object are unified—through the unity of self-consciousness that is mediated through the advance of culture.

When Wilhelm Scherer calls language "the image of our innermost being," he is presupposing the organization of the human body as its prototypal image.[13] Thus it should be theoretically possible to uncover the relationship between the brain's functions and those of the speech organs—"for," as Friedrich Albert Lange remarks, "with speech, we are dealing not only with the precise measurement of the pressure of the lips as they produce a *b* or a *p* and the fluent sequence of movements among the speech organs when forming a word that is difficult to pronounce. But speech must also mean something, and therefore a manifold of connections must proceed from the places where the word sounds are combined to the place where the sense impressions are combined. We can hardly imagine these connections otherwise than each particular sensation or particular impulse to muscular movement finding itself represented in a whole series of cells in the cerebral cortex, each of which has in turn its own particular connections."[14]

The findings yielded through comparison of the image with the prototypal organs alone will not enrich our knowledge of anatomy and physiology, though these findings do seem to indicate there may be conclusions to be drawn about the history of the brain's development and that of the speech organs, a developmental history that is coextensive with that of self-consciousness.

That this history of development in reality commences with the first tool produced by the human hand, and that, in continuing to construct, the human being is laying the rails along which he eventually comes to reflect upon himself—this is the basic idea on which our entire investigation rests.

We cannot conclude this section without once more emphasizing the fact that the first unremarkable hand tool, its indispensability only increasing over time, asserts, as a force of culture, its absolutely equal significance with, if not its superiority over, the myriad more complex tools that have come after it.

It is obvious, then, in connection with the domain we have just traversed, that of all of modern technology's greatest creations, there is one we may set above all others without exception, one that serves to demonstrate this idea immediately in a machinal way.

And so, disregarding the fact that the first hand tool pervades every stage of development, no matter what branch of industry, we are choosing to focus here on the two outermost endpoints of the track of human civilization—the apparently most menial, the emergence of the common hammer, and the obviously most impressive, the activity of the high-speed printing press. This way, we see a presumed contradiction dissolve in light of the fact that, in building and equipping a printer's shop, in each of the thousandfold acts of the craftsman's and the artist's hand, from breaking ground to the careful maintenance of delicate machine parts, the hammer is and will always remain the soul of manufacture, in every construct of the human hand. For the hammer, as the most natural continuation of the hand, is appointed for the universal exercise of the power invested in the hand; as such, it has more of a share in this universality than does any other tool.

It should come as no surprise, then, that our reflections on the hammer fall in line with our reflections on language—or on writing and its machinal perfection, to be more precise.

The latter, like anything that undergoes a course of development, warrants a review of its beginnings in order to be properly evaluated. At this point we will let a voice speak from the columns of an outstanding technical journal in New York, the *Scientific American*—for the article is so perfectly consistent with the basic principle of our investigation that I could hardly refrain from literally translating it and providing it in its entirety to the reader as follows:

"Few people," says Mr. J. Richards, "in witnessing the use of a hammer, or in using one themselves, ever think of it as an engine giving out tons of force, concentrating and applying power by functions which, if performed by another mechanism, would involve trains of gearing, levers, or screws; and that such a mechanism, if employed instead of hammers, must lack that important function of applying force in any direction that the will may direct.

"A simple hand hammer is, in the abstract, one of the most intricate of mechanical agents, that is, its action is more difficult to analyze than that of many complex machines involving trains of mechanism; but our familiarity with hammers makes us overlook this fact, and the hammer has even been denied a place among those mechanical contrivances to which there has been applied the mistaken name of mechanical powers.

"Let the reader compare a hammer with a wheel and axle, inclined plane, screw, or lever, as an agent for concentrating and applying power, noting the principles of its action first, and then considering its universal use, and he will conclude that if there is a mechanical device that comprehends distinct principles, that device is the common hammer; it seems, indeed, to be one of those things provided to meet a human necessity, and without which mechanical industry could not be carried on.

"In the manipulation of nearly every kind of material, the hammer is continually necessary in order to exert a force beyond what the hands may do, unaided by mechanism to multiply their force. A carpenter in driving a spike requires a force of from one to two tons, a blacksmith requires a force of from five pounds to five tons to meet the requirements of his work, a stonemason applies a force of from one hundred to one thousand pounds in driving the edge of his tools; chipping, calking, in fact nearly all mechanical operations consist more or less in blows, and blows are but the application of an accumulated force expended throughout a limited distance.

"Considered as a mechanical agent, the hammer concentrates the power of the arms and applies it in a manner that meets the requirements of the work. If great force is needed, a long swing and slow blows accomplish tons; if but little force is required, a short swing and rapid blows will serve, the degree of force being not only continually at control but the direction at which it is applied also. Another mechanism, if used instead of hammers to perform the same duty, would require to be a complicated machine, and act in but one direction or in one plane."[15]

If this is the language of an intelligible technology that directs our gaze backward and vindicates the hammer on the basis of empirical correctness, then let the hammer also be exalted by a higher truth in the idea that urges us forward, clad in myth:

In "Odin and Thor as the Expression of the Spirit of the German People," Felix Dahn shows that, as the god of thunder, Thor's ideal significance is that he is the patron god of agriculture and all of human civilization. But the tool that aids him in all his creation and work, the tool through which he becomes an expression of the spirit of the German people, is his mighty hammer. "The deed that his powerful arm performs is to batter and grind down the craggy, barren, desolate mountains with the blow of his unerring stone hammer, Mjölnir. He strikes the defiant heads of stone giants with shattering blows, gradually transforming the limestone, granite, basalt cliffs that would not yield to the plow into the fertile farmlands where one day golden grain will billow. This is how Thor becomes the god of human civilization: his *stone hammer* is not only a weapon in the war against the rock giants but it also serves peaceful ends. Encounter with the hammer consecrates the maiden as a bride and hallows with peace the threshold of the home. The hammer throw provides the measure for land settlement and partitioning. It drives fence posts into the ground. It batters together logs to form bridges. It determines and alters boundaries. Indeed, it consecrates the pyre onto which the hands of the pious lay the dead to rest."[16]

Thor's hammer is the idealized stone hammer of prehistory, which, in the hand of the primitive human being, accomplishes the miracle of the first work.

In myth we see shimmering the enduring truth of an inner or an outer experience. Mjönir's truth endures even today in the hammer-blows of groundbreaking ceremonies, symbolically alluding to the foundation of hearth and home, family and national community.

Under the sign of the hammer, then, we now enter the sphere of the state, in which, according to the prototypal image of the whole individual human organism, all of the brain's and the hand's activities interpenetrate.

CHAPTER 13

The State

With the different zones of the body's structure as our guide, we have traversed the vast domain of artifacts and are now convinced that in each case the latter's cultural-historical legitimacy is provided to it by the organism.

At each step of the way, without exception, we discovered these artifacts to be the after-images of various organic regions, until at last we crossed over the threshold into the organism as a whole, in the artifact of language.

Though we were unjustified in speaking of a railway-organism or a telegraph-organism, this did not prevent us from conditionally conceiving of a "language-organism." For, apart from its range and the other qualities associated with the material of language, a structure very closely related to the organism emerges through its mobility—or volubility—which is supported by grammatical-logical groupings and by inflection.

We have also come to know well enough the deeper meaning underpinning the word *Buchstabe*, revealed by tracing its origin. And since now we are able to see very clearly the significance of the corresponding words, *litera* and *literae*, we cannot help but claim the human organism itself as the original and authentic *universitas litterarum*. One and the same cache of letters, after all, may be rearranged in endless combination to produce any text, from the elementary school primer to the scientific compendium—and certainly this is not the result of having shaken them up in a mechanical manner but results instead from their having been conjugated according to an organically articulated rule.

Language is fundamentally distinct from all other formations of organ projection, in that it is not the image of an isolated group of organs, understood independently in its own right; rather, it is the image of a totality of functional organic relations. As the transparent form of an overall image of the organism, were language to remain thus abstracted from the technology that is also at its disposal, then in effect it would remain merely a specter of the organism. But we know that the development of language through technological means and the reverse, the development of technology by means of language, appear as the two sides of a single organic unity. Our depiction was perhaps unable to keep pace with this reciprocal, interpenetrative process involving brain and hand, and we were thus confined to representing this reciprocity in succession, first from one side, then the other.

Seen from this perspective, and especially since we are now moving from a discussion of language to a discussion of the state, an expression attributed to ancient Anacharsis—*man acts [handle] in accordance with words he thinks up himself*—acquires a curious new meaning.

We have so far variously devoted our attention to the ways in which the human being handles things, as well as to the letters and the language in which his thoughts, consciousness, and knowledge are formed and expressed. His actions, however, have remained somewhat remote. He is as responsible for these as he is for the specific activities he undertakes reflexively both in and toward society in general. The universal sphere of human responsibility, however, is the state.

Human activity, including artifactive activity (*hantalunga*, Old High German for the processing of something), becomes human action (*handelunge*, Middle High German for treatment, negotiation), becomes intentional conduct.

It does not matter how a hermit spends his time, since the human being has value as such only in human society. It is up to society alone to approve or disapprove of and to seek retributive action for the doings of an individual.

But in every human commonwealth, the one thing that all individuals have in common is the bodily human organism.[1] And so the state too is an evolving organism—it evolves, in other words, from the *res interna* of human nature and its total projection outward to become the *res externa*, the *res publica*.

Every natural organism is in the first place an end in itself. The end of the state, then, is none other than the uninhibited flow of its own spontaneous activity, its being-organism. The individual, though, is accountable to the purpose or the end of the state, whether he seeks to promote or to disturb it; his activities are acts.

In light of the following idea—that the individual, as an organic being [*Wesen*], cannot enter into contradiction with any outward extension of himself, such as the organism of state, without at the same time falling into contradiction with himself—we see that the individual is responsible for integrating his own will with the general will. This co-responsibility, to the ends of the individual organism and to the ends of the total state organism, is the wellspring of ethics and the realized legal order.

It follows that the whole human being is in the state and the entire state is in human beings. The human being, it is true, is the *zoon politikon*; but the state is a *polisma anthropikon*! It does not take the head or the hand alone but the entire human being to accomplish human acts.

Thus, the substance of the human being and all that proceeds from him is also the substance of the state. Because the state, as organism, *has* nothing but *is* everything that comes to be in it and through it, things like country, peoples, handicraft, art, science, ethics, religion are not ancillary or subsidiary matters or so-called ends of the so-called state alone. Rather, the substance itself in which a given end is effected is what has the state, has a hold on it and will not let it go; likewise, the substance in which the state is active and alive adheres to ends of its own. There are no dead, abstract ends.

The first tool was, as we saw above, the first work; and the first work was the true beginning of history. The original division of labor—a gendered division proceeding from marriage and then expanding outward from the family into the emergent professions—gradually became, once the first estate structure was established, the true hallmark of the organic life of the state.

Since time immemorial the state has been nothing other than the ever-evolving *organization of labor*. It is in this sense that we should evaluate the narrower meaning of a term that has since become a byword for social reform efforts.

The time is past in which comparison between the state and the bodily organism could be considered a merely figurative play of words. Many important discoveries have been made in the field of physiology that compel us to evaluate the comparison on more serious grounds. And how could it be the case that those we have already cited in support of a principle of organ projection would simply deny to the entire organism what they have already conceded as self-evident in the activity of the several substructures in its general organization?

Nevertheless, this way of thinking—subsuming the organism of the state under the unconscious prototypicality of the individual organism, meanwhile assuming knowledge of the latter through comparison with the former—is still far too uncommon, even in scientific circles, to have obviated the need for citation of further evidence from qualified sources, in order to eliminate doubts and misgivings.

Adolf Bastian introduces his work on *The Legal Relationships of Diverse Peoples* with the following words: "The organism of humankind, having the character of a *zoon politikon*, subdivides into its state institutions; as the lawful expression of a unitary existence, these institutions must manifest everywhere in the same broad outlines."[2]

In his well-known *Four Lectures on Life and Illness*, Virchow explains that no one is entitled to imagine the body as a machine that the soul steers according to its opinions, that instead the human body must be conceived of as a vigorous many-limbed organism. He compares the body with the family, the state, and society. "Here, too, the small and powerless stand beside the great and powerful, the common man beside magnates and potentates, each with his own life and essence that has its unique individual expression."[3]

"We can think of our own organism as a large state, as it were." This is how Max Perls begins his detailed comparison of the organs of the human body and their functions with those of the state.[4]

In his essay *Social Knowledge*, A. F. Grohmann undertakes to compare social formations among animals with the biological organism. He dismisses out of hand the accusation that such comparisons rely on mere simile. In defense of his position, he invokes John Tyndall, who for instance explained that it is no mere simile to call a glacier a current that heaves along a slow-moving mass; rather, as an analogy, the comparison presumes the equality of all defining circumstances. He then adds: "Were one to call human society a living organism, this too would be an analogy presuming the equality of all defining circumstances."[5]

Similarly, Walther Flemming, in *The Task of the Microscope Today*, calls the human body a great social republic: its cells represent the countless citizens, all of whom are equally enfranchised, though not all equally capable. Flemming too at several points elaborates the comparison in detail.[6]

In the first volume of his *Prehistory of Humankind*, Caspari deals with this issue at length.[7] For him, it is primarily a matter of clarifying the social question, to which end the natural history of cell life acts as a bridge. In animal swarms and herds, and especially the "state" of the hydromedusa, he locates the natural analogical link to the division of labor and the organization of society. One is hard pressed to imagine a social science today that does not seek counsel from the natural sciences. We are used to hearing economists and statesmen pronounce on the social question, but we would not be wrong in putting the question to physiologists and natural historians for a change. The well-formed, healthy organism, its structure and life, provide the prototypal image for a configuration of the social life of individuals within society, for, in effect, the organism represents a society in solidarity—that is, a confederated society of individual cells.

Gustav Jäger also elaborates the perfect analogy between the state and the natural organism by applying to animal forms terms like *republic, federated state,* and *constitutional monarchy.*[8]

Both Caspari and Jäger present the state as analogous with animal forms of lesser complexity, but this does not detract in the least from the cogency of their arguments. If these life-forms prove sufficient to the comparison, then the far more complex human body can only be all the more so.

Ernst Haeckel characterizes each higher organism as a society or a state composed of polymorphous elementary individuals, diversely developed through division of labor.[9]

Eduard von Hartmann added an appendix to the seventh edition of the first volume of his *Philosophy of the Unconscious,* titled "On the Physiology of the Nerve Centers." In it, he discusses in detail the correspondence between the natural organism and the organism of state, on the basis of important new discoveries about the physiology of the nervous system. "As an example of an elaborate system linking the governing elites, departmental regulation, local self-administration, and individual actors," he writes, the natural organism "occupies the middle ground between a democratic anarchy and a centralized prefectural economy."[10]

On the assumption that it would be rare these to days to encounter serious disagreements with any of the above-cited remarks, allow me to refer to a paper on the idea of the state which I published back in 1849.[11] I would hardly dare mention this if it were not for the sake of demonstrating that I do already have some experience in dealing with the subject. This paper originally evolved out of a reading of *Psyche* and took as its basis Carus's theory that the key to a true psychology could only be sought in the unconscious. Mine may have been the first thoroughgoing attempt at discerning in the human bodily organism the unconscious prototypal image of the organism of state. But critics, unable to appreciate that there might yet turn out to be a physiological basis for the analogical comparison, proceeded to label my investigation a prettily argued "simile."

The sole exception to this reception must have seemed all the more valuable to me then—a letter from Carus himself. I am excerpting here a passage from that letter that should be of some interest to more than just its recipient: "You have realized quite rightly that insights like these supply the cornerstone for many buildings that remain yet to be built, and that when it comes to the construction of the state in particular, the theory of the best constitution can only be enacted on the grounds of a higher and organic understanding. I have myself thought about these things in a variety of ways and recorded these thoughts; but to implement them—in such a way that they would be of immediate benefit to the life of the state—has presented difficulties. You rightly appreciate that what is essential—indeed foundational—is the theory of the unconscious in peoples and its evolution into consciousness in the law. As time goes on, there will be much for you to think further about and to change."

And indeed, I have had much to think further about and to change! The basic idea, however—that the state is an organism unconsciously proceeding from the work of the human hand and the human spirit, and that the bodily organism is the natural state—has remained firmly intact, especially since this idea was later affirmed by views such as those outlined above.

Incidentally—and this comes toward the end of his depiction of the "higher vital actions of our organism" in *Physis*, which was not published until after my own paper—Carus expressly beholds in these "the image of the whole of humankind, indeed the *archetypal image of its genuine political life*." He then adds, one might say boldly, that the statesman could stand to gain more insight, more of a sense of legal rationale, from the teachings of physiology than he can from bundles of old documents and parchments thick with dust.[12]

Under the circumstances, perhaps it will be opportune to provide a quick summary of at least the main points in my earlier conception of the idea of the state—which I still maintain is crucial.

Starting with the bodily organism as the unconscious prototypal image of the state, I attempted to discern where the idea of the state had its natural reservoirs, to throw into relief the fundamental natural law, and to clarify the relationship between the work of organization and the organization of labor. I went on to consider the state as the after-image of the bodily organism—of which we only gradually become conscious—by elaborating on popular representation, the monarchical principle, and the bureaucratic world. At the end came a discussion of the concept of freedom and its realization through social progress, which depends on the incremental expansion of consciousness, which in turn the human being achieves by referring the analogon of the state back to its prototypal image in the organic body.

This idea—that the state is the human society's consciousness, revealing itself in organic form—rests on the agreement between two aspects of the same fundamental law, the one governing nature and the other the state. If the agreement is sound, then we may observe the following:

> the *incalculable proliferation of the elementary organic form* into the greatest varieties of organization and, in terms of the state, the proliferation of the family in a manner similar to that which occurs in the process of cell formation;
>
> the *reiteration of smaller, more basic circles* of development, in increasing size and complexity, appearing in the state as administrative proceedings, from the district to the national level;
>
> the *law of periodicity*, recognizable in the periodic renewal and perpetuation of organic functions, and then in the total order of life in the state, which rests on a strict rule of periodic recurrence; and finally,

the truly conservative *law of metabolism* that maintains even the body of the state as unresting reform.

If we assume these laws are valid, then we must also recognize, in their interdependency, that they can only emanate from one and the same fundamental law of organic vitality. For each new emergence, each new organization of the organic complex through endless reiteration of a single elementary form—this is the progressive unfolding of ever higher life-circles, enabled through the self-actuating capacity of the organs to periodically expunge what is defunct and generate anew.

Now, just as the activity of an individual's limbs is comparable with the specialized work of an individual in society, the functions of the great integrated organic assemblage (digestive, circulatory, respiratory, and nervous systems) are comparable with the organization of the masses into professionalized estates (agriculture, industrial production, commerce, and intellectual work). And all of the following have their organic analogues as well: the diversification and outward expansion of the activities of professionalized estates as well as the foundation of parent companies and subsidiaries, colonialism as well as the existence of the armed forces for the protection of society. These analogues consist in part in the body parts required to propagate the species and in part in the skeletal framework, which, known by now as the organ of jointure and autonomy, amounts to the distinction of members and the bearing up of the whole.

There are just as many activities fundamental to the state as there are activities fundamental to the bodily organism. The state's subsistence, its stability, and its estates all depend on these fundamental activities. Nature presaged and pre-organized these activities for the state. Their upward reflex forms ministries. Were there to occur an oversupply in one or another sector of an estate activity, conflicting interests would cause the sector to branch; a corresponding branch would then also form at the highest levels of administration.

Thus: the foundation in which the commonwealth is rooted is estate labor, and the apexes in which it culminates are estate ministries. The general provincial representative bodies of old have since been consolidated into these ministries. Above and below, the complete estate! Still, between top and bottom, between foundation and cupola, disaffiliation and dissolution of estate solidarity does periodically occur, for the purpose of direct elections to the representative body of the people. Is this, then, renunciation and denial of the estate idea? Or is it a safeguard against an ossifying insularity? Is it the ruin or the rejuvenation of the idea of the state? The answer depends on a correct understanding of what we mean by mechanical degeneration and organic integration.

This, then, in broad strokes, was the essential content of a text whose task it was to discover the organic provenance of the idea of the state. To have proceeded any further with the comparisons it engages would have meant digressing, perhaps to the point of confusion. For, as long as the details of a question behave more like a wind-blown field than hard-packed ground, though enticing, that sort of ground is far too shaky and would jeopardize research prematurely.

Meanwhile, however, Albert Schäffle has taken up the subject on the basis of recent advances in the anthropological sciences and put together a wonderful work on *The Construction and Life of the Social Body*.[13] He calls it an encyclopedic outline for a real anatomy, physiology, and psychology of human society, with particular emphasis on the economy as social metabolism. He finds his main supports in basic anatomical and physiological facts concerning the organic body. Given the circumspection and exhaustive detail that is worked into this book, much of it borrowed from histology, it is ideally suited to securing long overdue recognition for the central idea that the social body is a self-organizing commonwealth on the model of the bodily organism.

It should by now be generally accepted that the state is an organism formed in the image of the human body. And since we know something of its provenance, we should be competent to posit something concerning its future. In order not to err, however, we must distinguish clearly between the indeterminate representation of the state, appearing in each present as historical fact, and the concept of the state, understood as a fully developed organism.

Representations—those forms belonging to history in which the idea of the state successively and simultaneously strives toward realization—are many; but the concept, the idea of the state, its aim exists only once. The former are temporal accretions to the state organism, in variable forms, each with its own particular national content; the latter is the singular concept of the organic, alive in each individual form of the state and shaping it. The former are changeable, evanescent; the latter, invariable and enduring.

According to the ideal view, which places the object under the auspices of the future, all states belonging to the past and to each unique present are, as factors of the highest organic development, means to an ultimate end. By contrast, political practice commonly mistakes the produce of its own machinations for the true state and erroneously declares what should be for it the unwavering guideline specifying its ultimate measure and aim instead to be the spawn of a sterile ideology.

It is true that practical routine causes the mechanism to drift and that sterile ideology spawns prematurely expiring formations. On the one hand, political stagnation; on the other, political precocity—both of them cliffs temporarily impeding the measured course of the state's development, under the unconscious direction of the organic idea. The development of the state, though, will never founder on either. For, human society is, by virtue of its becoming increasingly conscious, in the long run, of the laws of the unconscious that operate in the organism in the form of its legal constitution, safeguarded against the danger of mechanical devitalization and therefore capable of ascending to new and ever higher stages of self-consciousness.

The variable forms of the state must be recognized as means to the single end that is immanent in them. But if the state is regarded as the means to an end believed to lie somewhere beyond itself, then the question needs to be reformulated before it can be decided whether or not a given view of the state—either as means or as end or as an end in itself—is at all compatible with its nature. Franz Ziegler, celebrated patriot and distinguished for his statesmanlike insights, once lamented that "the idea of the state, of the greatest thing the human being is capable of creating," has been lost to us. Now, if the state is the greatest thing the human being is capable of creating, but only insofar as the state is an organism, then the idea of the state is in effect the idea of the organism, which ultimately means that the idea of the state is the greatest thing that human thought actually can attain. It should also be evident that we are not talking about the concept of the individual organic being as the norm. Rather, we are talking only of the organic idea, the being-organism, organicity—the supreme, divine life-source of every single organic structure.

Human thought—which is justified only within the *logos* and must therefore reject both fantasy and presentiment—has never yet presumed to have moved beyond this highest idea. The names of all those who basically amount to the main content of the philosophical construct are, properly speaking, merely predicates, their exclusive subject being, once and for all, the organic idea, the being-universal-organism. In this, we are and live and move. It is for this reason that the human being installs himself, life and limb, what he is and has, in the state. He creates nothing higher than the state; he thinks nothing higher than the organic idea; and he frankly does not extend, in body or mind, beyond the highest, the apex of all organic creation, which he himself is.

The organic idea is thus the very substance of the Absolute; it is the sole infallible grounds on which anthropomorphic attributes may be faithfully appropriated, in complete integrity. The organic idea is of cosmic universality. In it, thought also moves toward the microcosmic, the only grounds that never shift beneath it, because thought hews to the human being himself.

The human being, as the highest form of organic life, matures of itself into the life of the state, unconsciously in the beginning. For "the true constitution," according to Hegel, "lies in store for each people and each people strides toward it." So too all humanity strides toward the idea of the state, toward the ideal organism, composed of its many subordinated life-circles; humanity in effect strides toward itself. Along the way, and depending on which of the two reciprocally interactive factors—the conscious or the unconscious—predominates at any given moment in the development of self-consciousness, that which at one time had been an unconscious purpose or aim eventually becomes the other, conscious aim or purpose.

According to the concept of the organism, the state cannot be said to be either above or beside or in opposition to the people. If this nevertheless occurs, an internal contradiction will arise, based on the fact that under conditions of absolutism spontaneous activity is not evenly distributed throughout the whole; rather, it is effective from one side only and the other side therefore must experience it as coercion and domination.

On the other hand, if all citizens are more or less collectively engaged in the work of the state, then the work is common to all—it is truly a commonwealth [*Gemeinwesen*]. Its prototypal image is the self-governing, self-structuring, self-preserving human body. For what is characteristic of the organism is that it self-generates, such that what is produced is at the same time what is producing, similar to a process we have already seen at work in the organism of language. The state body [*Staatswesen*] too is self-generating and self-constituting; the state is its constitution. The constitution of the state happens on the analogy of that of the human body. The expression "constitution" ultimately has only these two meanings. Recalling the way in which the names given to human works are conveyed to their natural protoypal images, the expression is also evidence of the fact that the projection of a total organism—which the state is—also provides self-consciousness with a long view onto its own furthest reaches by retrospectively conveying information about total organic self-governance.

While there has lately been an increasing number of comparisons made between the state and the organism, there is also no shortage of attempts to delve deeper into the fact that our understanding of the natural organism could be furthered through reflexive meditation on the state that is unconsciously formed in its image, and that, therefore, the work of human beings could be used as a kind of physiological apparatus, entirely in keeping with the principle of organ projection.

It is with this in mind that Ernst Haeckel takes on the immensely difficult problem of the developmental history of humankind in *Anthropogeny*. He traces the problem back to the question: through what natural processes does that complex life-apparatus emerge, with all its diverse organs, from only a single cell? And he answers this question by remarking that the multicellular organism is structured and composed by the very same laws as is a civilized state, in which a variety of citizens are combined for a variety of endeavors and toward common ends. Haeckel assigns the greatest importance to precisely this comparison, because it easily facilitates our understanding of the composition of the human being from many heterogeneous cells and their seamless cooperation. "If we adhere to this comparison and apply this meaningful conception of the complex multicellular organism as a federated state to the history of its development, then we arrive at an important understanding of the actual nature of the first and most crucial developmental processes."[14]

We will have to forgo further extrapolation of the comparison. But this at least should be emphasized: that the comparison, insofar as it is considered "fundamental" to anthropogeny, is of the utmost importance, with regard to the priorities outlined here.

When Haeckel, speaking of an "understanding of the seamless co-operation of the different cells for an *apparently* premeditated purpose," emphasizes this "apparently," one may well assume that inclusion or omission of this one word represents the point of balance on which rests the whole "for" and "against" debate that has fervidly divided the scientific world.

Yet, with or without it, anthropogeny remains too solid a structure to be put out of joint by attacks in the usual way. To be sure, should the error of which he is accused indirectly redound to his favor, then Haeckel would in effect be his own most dangerous opponent.

For his prognosis of the inevitable bankruptcy of a spirit alleged to be thinking autonomously in matter would, considering the expenditure of spirit such prognosis entails, bear witness to the very opposite. Language and logic, with which the human being's original disposition endows him—which, however, to the strict materialist, are a chance mechanical supplement—must in this case prove to be the indomitable bulwark of the Genius striving toward a goal and conscious of the goal achieved. Haeckel's bold argumentation, the art of his drawings, his express enthusiasm for the impact he intends this work to have for his own premeditated aims—are these incidental and auxiliary to the organism? Or are they the expression of its interiority? And if it is the latter, then what do we make of an organic interiority whose expression betrays such explicit intent?

For our part, we adhere to Haeckel's acknowledged foundation of *Anthropogeny*, the "fundamental comparison" of the state and the human body; we further maintain that the human being becomes conscious of his own bodily life through knowledge of the state organism; and, in the interest of organ projection, we cannot but infer from this that every thinker is a bondman to science—quite literally owned in body—since the idea of a developmental history of humankind would not have simply arrived in the minds of the human being from somewhere outside of it. Rather, the researcher receives the idea from the object of his research, from the individual organism, which is none other than himself. The history of development is immanent in its idea; for the individual organism is the development idea enfleshed. Spirit is in the idea, and with spirit there is purpose in all development. Development can in no way be reduced to the predicate of laws of inheritance and adaptation. In fact it works the other way around, for development itself is the principal subject.

If what we have said does not yet fully satisfy the "fundamental" requirement in the comparison of the state and the human body, then all that remains at this point is to prove that the comparison also holds for the organic precursors to humankind. Caspari shows how this is done in his *Prehistory of Humankind,* in the section dealing with "the division of labor as the foundation and cause of all organization and of the organic life of the state." Among other examples, he offers the hydromedusa here as a model of the state, a kind of federative commonwealth, and on the whole loosely centralized organization in which the government is represented in the central polyp, the military estate in the tentacles for steering and capturing prey, the bureaucratic and intellectual estates in the touching and sensing tentacles, and the occupants of the interior, divided into male and female, represent the nutritional estate, which is responsible for production generally.[15]

Because actual empiricism must also be able to account for abnormal conditions in the organism, if we wish to fully substantiate the truly fundamental nature of the comparison by means of actual empiricism, then the lessons that pathology has always been able to learn from the diseased conditions of political society should be no less important for us as well.[16] And so, for instance, a political revolution resembles an acute fever, both being processes of healing that aim at ridding the organism of accumulated molt. In the state, such molt inhibits reform; in the human body, it inhibits metabolism. Therefore the need of reform has long been lamented: "The laws and statutes of a nation are an inherited disease, from generation unto generation . . . they drag on by degrees."[17]

It is therefore the task of the state to ward off mechanical disruptions and to maintain overall organic activity in an uninhibited flux. Mechanism is a drain on the organism, just as illness is a drain on health; mechanical deterioration and organic revitalization exist in inverse proportion. In the state, healing is achieved through work, but only through the kind of work that preserves and enhances vital powers, just as medicaments that are at the same time nutriments promise to do the most for the ailing body.

Conditions governed by a derelict principle of laissez-faire have always come to an end whenever work efforts, having relinquished complete organic solidarity and been allowed to run lawlessly to seed, fall victim to foreign persons and lands, to misery and defenselessness. This will continue to be the situation until the danger threatening the commonwealth brings the state to its senses and it begins to perceive the duty it has and the advantage it gains in the taming of the organic environment. Obviously, there is no one special kind of work that alone promotes healing, for instance. In the organism—which consists in the fact that the form and function of the organs are identical—everything is work. None of its life-circles alone guarantees life but, in working, each guarantees the others. Today's political economy still has much to learn from ancient Menenius Agrippa and his so-called allegory *The Belly and Its Members*. As far as we know, he was the first in his day to formulate the fundamental comparison as *argumentum ad plebem*, which, along with the biblical version, "And whether one member suffer, all the members suffer with it," has remained foundational, both in terms of political wisdom and in terms of the art of the individual life.

There is no lack of remarkable historical examples showing the state gradually overcoming the difficulties mentioned above. These types of difficulties were all the greater at a time when unruly and noncompliant private syndicates were permitted to operate within society in scattered territorial possessions, which fostered a kind of sequestration.

Our civilized nations today face a similar exigency.

By way of explanation, recall that "in the Middle Ages, the religious chivalric orders represented the military estate, in absence of a standing army, though they were sequestered from the state community. Likewise the monastic orders represented the church, the Vehmic courts the judiciary, and the Hanseatic League commerce. These secular and spiritual corporations frequently overlapped into one another's purviews and into that of the state. The clergy also carried the sword, merchants handled their own jurisdiction and warfare, and so on. Given these kinds of political frictions, the corporations could not but have fallen afoul of the state with time, for the state, in order to be equal to its concept, had to exercise these vital functions on its own. It was just as unlikely that the corporations would escape insertion into a higher order as it was that their occasional subsidiaries would escape incorporation into the larger whole."[18] It was a while yet before schools grew out of monasteries and commercial activity out of guilds; before art and science were housed in academies and universities, under government patronage; before special ministries of agriculture, trade, and social change were placed in the context of a national economy; before at long last monopolies and special privileges, usurious parasites on the marrow of the state, gave way to free competition.

It is clear from the entire course of these events that, as long as the state's collective work effort returns to its own organic proportion, the mass instinct will initiate on its own whatever aid is needed. What's also clear is that it is only after the child reared in the private sector has, in the throes of power, misappropriated the commonwealth that the state can successfully accomplish the work of its own self-organization by organizing the work efforts of all of its citizens.

The words "change" [*Wandel*] and "exchange" [*Handel*] may be associated in a neat expression to describe general economic activity. The expression would refer above all to roads and all forms of means of communication, since these are what enable change and exchange—that is, the transformation and reconfiguration of raw products as well as the migration of goods and persons. Foremost among these means is the post.

We will focus on the post in particular, in connection with our earlier discussion of conduction by rail and wire and in the context of the development of the organism of state. We also have another motive for this— that being the coalition presently in effect in Germany between the state telegraph administration and the office of the postmaster general.

A general remark about the relationship of the post to railways and telegraphs has to do with what we call them. One commonly speaks of railway systems and telegraph systems, though this is seldom the case with the post, which is a service and also an office [*Postwesen*].

The difference in these terms can be explained as follows: the mechanical apparatuses for the movement of telegrams and locomotives are permanently bound in fixed relationships along unchanging lines made of wires and rails, whereas the post employs a very versatile and independent set of technical means. After having been more or less provisionally established on rail lines and telegraphs, the post is now on the brink of permanently incorporating these structures to itself. Letters have long since become telegrams, and telegrams letters. Country roads,

postal vans, and letters—these undeveloped forms of rail lines, locomotives, and telegrams—nevertheless continue to serve as means of transport alongside steam and electricity. The post's freight and passenger vans
supply railroad cars, the semaphore telegraph is now in service to the
railway, such that what formerly had been an end becomes yet another
factor in attaining a new and better space- and time-abbreviating end.

The concept of the post is inclusive of all means of transportation
in service to the state—from the pigeon to the quadruped, the bicycle
to the locomotive, the letter carrier to the courier, the meadow path to
the iron rail, the town crier to the pneumatic dispatch, plus all the tubes,
wires, cables, tunnels, and steamers moving through the air or over or
under or by land and water.

The post is, in a real sense, the transportation of the means of state
by means of the state. The state must, finally, absorb all of the functions
belonging to it, and this it accomplishes by relieving of their isolation those
formations that have maintained their viability within its sphere, by setting
these and the institutions of public commerce together in an organic flux.

As a state institution, the post is locomotion parlayed into global commerce. Its former concept has become too narrow. The idea of a global post and its innovative implementation is the very inauguration of
the commerce ministry of the future, an operation of universal proportion, encompassing the entire civilized world. In terms of post, one
should envision not so much the external technical material of locomotion but its purpose, its inner tendency, and its ramification. The post
is steeped in the essence and substance of the state; it is the organic life
of a national spirit. By contrast, it is the piecemeal manner in which
technical apparatuses are assembled that clearly marks the telegraph and
railways as mechanical compositions; we notice this in the characteristic
monotony of their structure, and it is this that established them as systems, in the terminology of industry.

That said, the post is the state form of communication and, depending
on its degree of perfection, the image of the unbroken correspondence
of all the functions in the individual organism. As such, it is well suited
to convey the correct representation of organic vitality, for the sake of
advancing self-consciousness!

In the conversation between Goethe and Napoleon at Erfurt, their talk
came around to the topic of the ancients' idea of fate. It was remarked

that in the modern era it is the power of circumstances that has come to stand in for fate. But what is the power of circumstances? Drawing on the basic premise of our investigation, the circumstances holding the human being in their power are, to his advantage, his own cultural conditions, the product of the work of his hand and spirit. All other creatures are subject primarily to the power of natural events. Circumstances, however, presuppose work and consciousness. One tends not to speak of the favorable or unfavorable circumstances of plants or animals. The fact that elementary forces intervene decisively in the shaping of culture, as do human will and intelligence in the shaping of nature, ultimately makes little difference, because what remains fundamental to the former are culture and the state community, to the latter nature and the unconscious life of the herd.

Now, the human body participates in the composition of the cultural sphere in such a way that the various entities of organ projection—which are inconceivable outside the context of an organic whole—are necessarily unconsciously affected by the idea their makers have of this whole. It is inevitable, then, that this infusion of the organic in them will aid in realizing the only sphere within which their emergence and perfection is even possible—that is, the state and its development. For there has never been a stateless culture nor a cultureless state.

It is therefore from the identity of culture and state that the power of circumstances arises. Commerce, justice, arms, police, schools, art and science, the church, posts, roads, and telegraphs are all administered by this fate. And the railways?[19]

How the particular functions in the total life of the state are ordered will vary greatly, depending on the degree of need. Since the distance to the goal—that being the perfect harmony of action and reaction—is always infinite, the most important step will already have been taken once resistance to the organizational right of the state is broken and the object is under its control.

Depending on the power of circumstances, it is possible that one or another aspect of organic vitality may come to the fore, to a certain extent. In such circumstances, one distinguishes among agricultural, commercial, industrial, clerical, and military states. Should national pathos emerge strongly in one of these aspects, then the strength of this feeling is essentially marked by the fact that the less developed aspects will seek protection and prosperity in it. In this form, the latter recover what they invest in working for the subsistence of the whole, in accordance with the organic principle of reciprocal exchange. The self-preservation of the organism is based on this, in general; so too is the superiority of the German military constitution, in particular. History has yet to present us with a more highly developed form of organization of the functions of state than this.

Imbued as the army is by the above-mentioned fundamental natural law, we recognize in the individual troops the elementary forms of tactical organization; therein, we find lower spheres reiterated in ever higher spheres, all the way up to the general staff; this process of organization is regulated through a series of periodically recurring drills and by the metabolic exchange of both material and power through recruitment and transfer of staff. This ensures, to the greatest possible extent, that a mechanical rust keeps from forming and that the army may then proceed along the correct path toward its eminent goal of becoming the prototype for the full organic development of the remaining estates.

The rigor of military order and subordination is an outgrowth of the immutable fundamental natural law. This rigor has also mistakenly earned the German army, the only one under discussion here, the label of militarism, an expression that implies aversion to the same. This epithet is often heard in the context of talk about the "abolition" of the standing army; what abolitionists, however, fail to consider is the fact that bands of mercenaries may be bought and sold but the limbs of an organism are inherent and grow from the inside out.

An army organization, which has its roots in folk-traditional military service and in the national disposition, emerges and gradually grows to meet its own goals—which are also the unconscious goals of the nation.

It is only technical armament and deployment for a specific action that happen with conscious intention. And yet: all those who took part in founding the army, who of them could even have imagined the extent of its development today!

However, in lockstep with great scientific discoveries, we see this development periodically disrupted by conscious acts of variable energy and duration.

Ever since the last general draft, Germany has been in the midst of a period of reflection on the army's past training and efficiency and the role it will have in future. The belief is gaining ground that such a highly developed outgrowth of the national consciousness—which has itself grown conjointly with the organism of state—cannot be simply decreed away without the whole organism collapsing.

Even the demand for partial disarmament will ultimately have to be decided based on the scientific explanation of reflex motion—since to increase the military budget is a political reflex. There is no appeal with organic laws; they are themselves the highest authority.

In the face of unreasonable expectations of disarmament, demobilization, demilitarization, and the complete abolition of the army, a further insight surfaces: the standing army, apart from its obvious occupation of defending the state's political autonomy on the battlefield, also has a share in safeguarding science.

For the art of war has since transformed into a highly developed military science and, as such, is affiliated with the other sciences, nourished by them and contributing to their advance. Science, though, grows out of school, which includes all instructional and educational institutions, from kindergarten to university. The military schools sprung from this common root have gained in strength to form their own vigorous branch. Scholarly and military institutions, looking on one another as siblings, are now en route to a probable merger of their joint interests. Representing state physical education on the one hand and suffering none of the infirmities that scholarly schools do on the other, the standing army indicates the goal of its education to be the overall balanced development of the human being's physical and intellectual dispositions. Hegel's saying, much criticized at the time, that the army belongs to the intellectual estate proves increasingly true.

The *exceptionally* privileged status of the army today is the natural consequence of its constant drilling in preparation for the *state of exception* of military conflict—which, when it happens, if the sovereignty of the state is at stake, will demand all the resilience of the nation that is gathered within the standing army. Thus the necessity of its rigorously imposed organization and thus the reason that as yet unmobilized forces instinctively gravitate toward it.

If university lecture halls were to close for an extended period of time, this would not endanger the existence of the state. By contrast, were the entire army to be discharged, even for a short while, this would unleash all the forces hostile to the state, both within and without it, unsettling it in its foundations. And to replace the standing army with a volunteer one would be tantamount to financial collapse and society's undermining.

By contrast, the constant, routine, commissioned service of the standing army functions as capital that is distributed unnoticed throughout the veins and arteries of the commonwealth as strength and health, obedience and a sense of duty. Its value, incalculable in and of itself, evades any roughly rational assessment for the unthinking habit of a natural daily pleasure.

Were one to ask what the generative power is that effects such activity, the answer could only be found in the cardinal relation between command and subordination. It is, after all, subordination to a higher will, beginning with the family, that engenders solidarity in a society. Straitened into irrevocable command and steadfast obedience, discipline—which is military spirit become substance of thought—should keep the army ready to commit itself unconditionally to the single command issued from above when the decisive moment arrives.

Now, in answer to the question concerning the connection between military discipline and the theme of our investigation, we will have to return to an earlier discussion.[20]

Above we saw that the development of the machine begins with the paired element. The characteristic features of the paired element are as follows: first, that one of the pair is the envelope-form of the other, and second, that their movement can only proceed in a manner unique to their pairing. If one element of the pair is fixed, then the movement of the other is absolute with respect to that fixed point; if neither element is fixed, the elements move relative to one another.

If the envelopment of one by the other is complete, the form of their closure is full; if the envelopment is only partial, this we called part-closure. Since the compulsory movement of the paired element with a part-closure is never entirely certain, whereas full-closure prevents any interference, the machine's ultimate perfection is based on the eventual supersession of part-closure by full-closure.

It is a truly surprising fact that the development of human society also has the same foundation—this being the subordination of the individual will to a higher one. The most prominent form of this is military discipline.

That said, we will call the part- and full-closure of the paired element the discipline of the machine, and the discipline of the army is accordingly the full-closure of command and obedience. We should note, moreover, the dual meaning of the term "discipline": on the one hand, the paired compliance of the elements determining the standing orders of the machine and of the army; on the other hand, the knowledge and science of the production and upkeep of these outcomes.

The fixed point or superior commanding will around which the obedient must move in paired closure, in order of rank descending from the top—this is the monarchic principle. This basic feature is present in even the most inadequately disciplined military organization.

Where we see steady progress in the transition from an imperfect to a perfect interlocking of command and execution, wherever the indeterminate is replaced by the determinate, the slack by the fixed, the exception by the rule, there we locate the grounds for the prestige of the German standing army. It stands at the highest stage of organization yet to have been achieved and is recognized as the model for every other military constitution. Its astonishing success is due to having raised blind obedience to an unconditional one, a result of its clear insight into the necessity of such for the survival of the whole. This inner necessity is the ground of human freedom, whereas the machine is regulated entirely by the coercive power of external necessity.

Even were there to be a period of time without warfare, the army would nevertheless remain indispensable—as an institution interwoven with the state and as the renewing grounds of obedience and physical efficiency. The simple fact of its existence holds anarchic impulses at bay.

The ways in which the army may go on to reconfigure itself, as culture in general continues to progress, are as unforeseeable to us at present as the present situation would have been to earlier generations. What is certain is that the army, unconsciously following the course of the idea of the state, must go on to discover its task of representing the strength of the commonwealth for as long as we continue to seek the totality of the personality in the balanced cultivation of human dispositions. It is too early, in any case, to agonize over tomorrow's canons and torpedoes.

If machine technology is indeed recognized as the germ and focal point of the sphere of mechanical constructs, and if the organization of the army is indeed understood as the prototypal image for further development of state society, then we have done more than merely suggest that there exists a crucial link between the two vast domains the human being creates—the machinal and the political. We have then actually confirmed that the development of both the state and the machine are based on the same principle. The machine, too, must obey!

In the standing army, the state has created an organ that vitally sustains, in its own discipline, the idea of obedience in society as a whole. Therefore, if we let our reflections linger just a while longer on this particular profession, it is because reasons arising directly from the inner nature of the state compel us to do so.

We need not continue emphasizing the importance of mechanical technology for the development of the state idea. We believe by now to have eliminated any doubt that, on account of their correlation, we gain a higher understanding of the state through the machine and of the machine through the state. Indeed, they share a common prototypal image in one and the same archetypal image—disciplinary full-closure! The one manner of movement that is permitted on the basis of paired congruence in the sphere of technology and in that of the state becomes, in the bodily organ, its purpose, which is at the same time its sole determination. Thus seeing is the purpose and sole determination of the eye. Thus too: in the machine, neither paired element may adopt another form of pair closure that is in any way contrary to its original kinematic form; likewise in the world of ethics, one may not expect the obedient to be able to satisfy two contrary commands at the same time.

In *The Kinematics of Machinery* it says, regarding the importance of the mutual relationship between political and machine science: "The whole inner nature of the machine is the result of a systematic restriction; its completeness indicates the artful constraint of movement to such point that any indeterminacy is entirely eliminated. Mankind has worked for ages in developing this limitation. If we look for a parallel to it elsewhere, we may find it in the great problem of human civilization. The development of machinery forms indeed but one factor in civilization, while at the same time the former presents us with an intensified counter-image of latter."[21]

Seldom is such a bright beam of knowledge cast upon the innermost nature of the life of the state. Originating in the sphere wherein, mechanically, "all is revolving," here we find a clue to the world of ethics wherein, organically, "all is flux." In both spheres: the artful constraint of movement. Only the elevated mood, into which a reader is plunged at the conclusion of this masterful work, could supply words of such simplicity and quiet grandeur!

Along the whole length of the path we have traveled, we were at pains to maintain an opposition between mechanical apparatuses and organic structures.

Meanwhile, the antithesis between a mechanical manufact and an organic factor has essentially been mitigated, since the machine has given way to another product of human labor—namely, the state. Suddenly we see the machine mechanism and the state organism moving "hand in hand" with the desire that science will disclose proof of their common origin.

Our preoccupation with artifactual objects in the world exterior to the human being played out in such a one-sided way that we did not devote equal attention to the concomitant emergence of the state, nor to the reciprocal conditionality of their joint realization. Now, though, we can no longer avoid the fact that the domain of mechanical formations—the isolated forms of which cannot themselves escape stark contrast with the bodily organism—actually and integrally belongs to the organism of state, in a single whole.

Now then, suddenly, we find we are dealing with two organic domains—with the individual human organism and the total organism of the state. The former stood face to face with isolated artifacts, mechanical contrivances in a variety of organizational forms and patterned after specific functional relationships—the human being on one side, the machine on the other; while the latter, the total organism of the state, discovered its archetypal image in the human body. But if one wished to maintain a distinction here—with the human being on one side and the state on the other—then the state, just like those isolated artifacts, would have to be adjudged a mechanism. For this reason, prototypicality alone does not suffice in this case. In point of fact, all individual organisms, in their complete *sui generis* corporeality, constitute the actual organic substance of the state. The prototypal image for the organization of their overall work, however, is the very same as that which unconsciously determined the manufacture of individual artifacts.

Just like the individual organism, whose originally inorganic or chemical components are recombined as a result of the organic idea in new ways that would not be possible in the field of chemistry alone, so too the material composition of the state, its land and all its technical contents, acquires a significance by participating in organic life that essentially distinguishes it from everything else that has never encountered human purposes.

Whenever the historical human being—that is, the human being living in state society—has impressed an object with the touch of his spirit, such stuff appears as taking part in history and, because the historical process is identical with the organism of state, this stuff forms a state-organic compound.

The living human body cannot exist without sensuous reality; nor is there a state body without sensuous matter from which it assembles itself. The totality of matter formed by the human hand is thus, as the total inventory of state-organic compounds, devoid of a mechanical character, just like the inorganic matter in the living body, ordered according to the organic idea, becomes its organic constitution.

The state, no matter if it is still imperfect or even deteriorating, remains an organism and is never a machine. Whether applied to an individual or to society as a whole, the term "machinelike" for the most part simply indicates a high degree of mindlessness and habitual routine. To be a state means to behave like an organism. This is why the state can never be entirely mechanical—though it does have its machines within it, which must, as individual mechanisms, be distinguished from individual organisms, but that are, as wholes within the whole, the very material the state itself makes fit to be appropriated for its own self-preservation.

We have now arrived at the summit of our investigation. Whereas earlier, the products of the human hand, in the form of individual mechanisms, were not to be confused with organic structures, now we see them begin to coalesce, in their totality, with the totality of human individuals into a social-organic unity. The antithesis that persists in the individual artifacts, the antithesis of mechanism and organism, is therefore sublated [*aufgehoben*] in this form, in the material store of the state body.

We placed all individual works patterned after an organic prototypal image under the concept of tool, to the extent that they served as means to understanding organic corporeality and to evolving self-consciousness. We also recognized spoken language as a tool, even though the hand did not participate directly in its development. Language, we found, is a tool that, as the product of its own activity, bears in itself the unmistakable mark of the organic, the immediate unity of the bringing forth and the brought forth. But language does require the hand's support for its development in the long run; and so, in handwriting and printed matter, language in a sense took a step backward into the manufactive development of tools. In this way—with one foot, as it were, in the sphere of the mechanism and the other in the sphere of the organic and of spirit; through sound and the content of thought, on the one hand, and through handwriting and printed matter on the other—language is the nexus of the two great creations of humankind, the material-mechanical world and the organic-spiritual world.

Their unity is realized in the state, in the reflection of a unity of the sensual and the spiritual appearing as the human being. The Hellenes understood the fullness of power in this unity by the word *energeia*—that is, generative work-activity, from which came the first blow of the hammer that announced the construction of state society.

We should realize that any given state can only act as partial evidence for our reflections. We should therefore have to extrapolate from all states that have perished and those still existing, as developmental forms of state, what the perfect form could be in the future and endeavor to gather the true measure from the idea of the state itself.

Were we to even roughly achieve this, we may assume we have arrived at our goal while at the same time find we are back again at the beginning—for the primitive human being of the time before tools with whom we began was the as yet still undisclosed organic vitality, the unity harboring within itself, immediately and unconsciously, a hand and brain and the whole civilized world.

From the original disposition of this primitive human being emerged weapons, tools, equipment, apparatuses, instruments, machines, and every alteration made to the face of the earth in order to support massive systems of exchange. Over the course of eons, like glacial masses flowing on imperceptibly, ever emerging and resubmerging in the flux of the organic idea, all of these return, finally, to the unity that is realized in the body of state. The end and the beginning are the same—the latter the untapped unity, the former the unity brimming with the fruits of human labor.

Again, we reiterate that conscious human activity is work; that the work performed within the state or anywhere for the sake of the state appears as estate labor; and that estate labor becomes action, assuming either a sustaining or disruptive character in relation to the state organism. For each of his acts—whether they have to do with forming matter or cultivating spirit, whether he constructs boilers or "infernal machines," attends public school or studies esoterica—the individual is accountable to society. This is why—with respect to the expression *man acts in accordance with words he thinks up himself* or, in other words, he acts deliberately—we have called the state the sphere of human responsibility. Whether this is moral or ethical responsibility depends on whether his actions affect the legal sphere of society, which is mutable, or the sovereign right of the organism and the conscience, which are eternal and immutable.

If, out of the depths of its unconscious intertwinement with state society, the human being becomes conscious of disruptions and injuries having occurred to what we referred to above as the corresponsibility of the individual organism's purposes and the purposes of the total organism of state, then, in having established that something is that should not be, the voice of conscience makes itself heard. The stirrings of conscience are acts of consciousness that, when repeated, shrink the boundaries of the unconscious in like degree as the boundaries of self-consciousness expand.

We have now reached the outer limit of organ projection, the limit beyond which sensibly perceptible, numerable, and measurable products cease to appear. We had to relinquish even the state as a means to our understanding of the body's substance. That being said, there is yet one further step to venture before the last glimmer of means and tool disappear entirely.

We are used to designating as "development" the essential trait in the life and limb, the configuration and reconfiguration, the coming to be and passing away, the growth and decay, even the unity of all specialized functions in the organism. In this immediate unity, development is to begin with purely a change in dimension. But then, projected in the distinction between dimension and change, development discovers a cosmic accord between the juxtaposition of things and the succession of their movements. The human being names this accord space and time and these provide him with the forms for his intuition of the phenomenal world. By means of these abstractions he is able to differentiate the world into spatiality and temporality.

One of humankind's greatest acts was to have released the concepts of space and time and extended them out into infinity and eternity. These concepts originated in the organic Self—in life and limb, in matter and motion, in dimension and change. By referring back to this origin of the thought-entities [*Gedankenwesen*] it has released, the Self maintains a contiguity, far surpassing the spatial and temporal, with what we referred to above as the Subject of the eternal and the infinite.

Our investigation has now arrived at its destination. Having accompanied the human being from his very first act of work, heralded in the tool, to the full-scale practice of professionalized estate labor, we have discerned human society coalescing, with all its cultural accoutrements, into a single organic unity in the body of the state. If we attempt to look beyond this body of state, we are not thereby abandoning the fixed standpoint that we already assumed in the formation of a state bearing a particular spatiality and temporality. Rather, we have simply left to the state idea and its special forms the right it already possesses to pull toward the panorganic source. To draw any further conclusions would be to exceed the scope of our intention—though we will suggest a couple of directions in which the theory of organ projection may be able to provide science with new perspectives from which to begin answering some of its questions.

Humankind is still in its cultural infancy, or rather at the start of the technological tracks that spirit will have to go on laying in order to advance. Nevertheless, even the comparatively short distance we have already come and what we have already achieved—with the increasing perfection of our tools and machines and by means of more adroit exploitation of natural forces—is enough to permit us to make bold inferences as to the future of civilization, the magnitude of which is beyond measure.

Corresponding to the increase in knowledge of the bodily organism, a higher self-consciousness will come to permeate society, which will have a curative effect. With the amelioration of conflict, of individuals amongst themselves and of states with one another, the pessimistic worldview will be reduced to the degree necessary for a healthy optimism to thrive.

Expanded and transformed into fixed thought content, knowledge of the basic organic conditions of the life of the state will simplify the law into the content and form of an ethical statute.

The apostolic message, which declared the body a temple of the holy spirit, will find its thus far fragmentary operations supplemented by a religion of the body—thereupon reconciling the demands of earthly reality with the injunction toward a transcendental beyond. The message will thus accomplish its self-renewal and deliver us from the evil of social grievances.

Knowledge beginning in the bodily self will yield greater information about as yet still unresolved problems of knowledge generally; about the provenance of the concepts space and time; about the thing-in-itself amid the multiplicity of things and its stepwise unveiling, occuring in relation with the development of self-consciousness.

Art will go on—intertwined with the source of all beauty, with the organic idea—to elevate the fundamental morphological law into a power capable of transfiguring all of life's circumstances.

We will find the insight corroborated that human freedom is predicated on the human being's capacity to freedom, the development of the latter keeping pace with the development of self-consciousness. Any tendency other than those toward which original human dispositions naturally strive imposes coercion and unfreedom on the will from without. It turns out, therefore, that true freedom is found only where there is a consciously willed, disciplined, and cultivated harmony with inner organic necessity. It also turns out that a spontaneous free will is necessary for the human being in order that he may attach value to his becoming what he should be. The will is conceivable only as housed inside the human being, beyond which it founders, on hubris above, on brutality below.

Because it presupposes the identity of freedom and inner necessity—as this emerges from our understanding of tool and machine, of the human being and work, according to the development of one from the other, the explanation of one through the other—the state gives an indication of our progressively higher conception of the organic idea. The success of all scientific endeavor rests on this idea; all philosophical systems move in it; it is the promise of a religious rebirth on the grounds of the one belief under whose symbol stand all the *civilized peoples* on earth.

And so that which is nearest, most obvious, most given to us, that from which all speculation departs and to which it returns, remains the human being—and his greatest real structure, the state, his own all in all.

From the tools and machines he has created, from the words he has thought up, the human being steps forth, the *deus ex machina*, face to face with himself!

A Media-Archaeological Postscript to the Translation of Ernst Kapp's *Elements of a Philosophy of Technology* (1877)

SIEGFRIED ZIELINSKI

Das Irrsein ist nichts anderes als das Ohne-Organ
. . . das Irrsein hat kein eigenes Organ
steigt alle Leben hinunter
irreseiend überall
Madness is nothing but the being-without-organ
. . . madness has no organ of its own
but descends after all life
madly everywhere

—Constanze Schwartzlin-Berberat, Waldau (ca. 1900)

PRAISE FOR THE TRANSLATIONS

Georges Canguilhem began writing his now legendary essay "Machine and Organism" while teaching philosophy in Strasbourg. The essay was published in 1952. In order to make use of Ernst Kapp's contribution toward a philosophy of technology, Canguilhem had to rely on Alfred Espinas's *Les origines de la technologie* rather than Kapp's original. Espinas's classic of the French humanities was published in 1897 in the series Étude Sociologique, twenty years after the pedagogue, geographer, and philosopher from the German province published his *Grundlinien einer Philosophie der Technik*. Espinas was among the first in France to acknowledge Kapp's work. He dedicated a section of the second chapter of his own book to discussing Kapp's intellectual master stroke, the "projection organique des mesures spatiales."

Seventy-five years after its first publication, Kapp's text remained unavailable to Canguilhem in French.[1] Two of Canguilhem's students, Gilbert Simondon and Michel Foucault, knew of Kapp not from the original but through the intermediaries of Canguilhem and Espinas. Many who were then working interdisciplinarily in France, exploring the relations between thinking and making, between philosophy and techno-poiesis, persisted in ascribing the innovative idea of organ projection to Espinas. Canguilhem, summarizing the idea, refers to Espinas's reading of Kapp rather than to Kapp himself when he writes: "According to the theory of projection, the philosophical foundations of which go all the way back, via Eduard von Hartmann's *Philosophy of the Unconscious*, to Schopenhauer, the earliest tools are nothing but extensions of movable human organs."[2] In any event, Canguilhem, along with Espinas, emphasizes that Kapp had made the prosthetic dimension of organ projection valid solely for the "earliest tools" and not for all conceivable instances of technology, arguing that the closer Kapp came to the technical reality of his own day—that is, to the complex evolving system of electric and networked objects—the more independent of organic structures the technical objects became, which were instead scaled to human consciousness.

Similarly to the role Espinas's reading of Kapp played in postwar French thought, the conservative German philosopher Arnold Gehlen's reading affected the Anglo-Saxon discussion. Gehlen's 1940 book *Der Mensch: Seine Natur und seine Stellung in der Welt* was translated into English as *Man: His Nature and Place in the World* (1988), as was his 1957 book *Die Seele im technischen Zeitalter*—the title a play on Günther Anders's epochal media-critical work *Die Antiquiertheit des Menschen*, the first volume of which was published in Munich in 1956 as *Über die Seele im Zeitalter der zweiten industriellen Revolution*. In English translation, Gehlen's later book appeared as *Man in the Age of Technology* (1980). Gehlen's Promethean man—the human predator, more or less underdeveloped relative to other mammals, who creates technologies for himself in order to live and survive as a deficient being—was influenced by the writings of Paul Alsberg and José Ortega y Gasset as well as Kapp's inquiries into the philosophy of technology, and was often accepted in lieu of the inventor of *organ projection*.

Carl Mitcham and Robert Mackey pioneered the introduction of European philosophies of science and technology in the United States. In their 1972 edited collection *Philosophy and Technology* they presented Kapp's book—albeit mediated through the writings of another[3]—as a crucial element of a systematic philosophy of technology. Mitcham, in his later study *Thinking through Technology: The Path between Engineering and Philosophy* (1994), dedicated a subchapter to Kapp's theory of organ projection. The philosophical master of metaphorology, Hans Blumenberg (1920–96), has been extensively translated into English; he refers time and again to the early champions of anthropological theories of technology, as does the influential (at least in the European context) philosopher and historian of technology Alois Huning in, for instance, his specialized study on the *Philosophy of Technology and the Verein Deutscher Ingenieure* (1990). One of the more recent publications that prominently features a discussion of Kapp is Fabio Grigenti's *Existence and Machine: The German Philosophy in the Age of Machines (1870–1960).*[4]

Odd displacements and circuitous cross references like these are no longer necessary—at least not so far as concerns the potential intellectual presence of Kapp's thought in Anglo-American, the contemporary Esperanto of the modern humanities. Thanks to the present complete English translation from the University of Minnesota Press, the provocative and contentious ideas of the German Hegelian and freethinker can now be assessed, 140 years on, on account of this transmission of the original textual material.

SOME PROVENANCES

With a view to the immanent genealogy of Kapp's work, it is necessary to point out an important context. Johannes Rohbeck has already discussed this in a powerful essay on Kapp's philosophy of technology. Under the influence of his most significant teacher, Carl Ritter, one of the founders of scientific geography, Kapp, a passionate instructor of history and geography, understood history in a high-modern sense—that is, as a motivated "geography in motion." His *Philosophische oder vergleichende allgemeine Erdkunde* (Philosophical or comparative universal geography), which was published in 1845–46, dealt, among other things, with the "influence of natural conditions on political, military and scientific formations" and early on unabashedly addressed a "culture of the present (means of

transportation and communication via 'universal telegraphics')" as well
as a "transfiguration of nature" through "work purified into art"—which
Rohbeck in turn affiliates with the tradition of the Hegelian idea of a
"unity of nature and spirit, of earth and man, mediated through the work
of human beings."[5]

Thirty years later, following his experiences living and working in Texas,
Kapp wrote his *Philosophie der Technik*, in which he logically pursues this
same idea, now projecting it onto a vast array of technical phenomena.
In this late work, Kapp synthesizes ideas regarding the interdependence
of philosophy and technology that had been floated and circulating for
centuries, culminating in a unique methodology at once constructive
and critical of civilization—that is, a philosophical anthropology of technology and its possible relations to culture.

In the archaeology of the philosophy of technology, two foundational
works demarcate the poles spanning the first century and a half of (natural) philosophical thought concerning technology: the ten books on
architecture *(De architectura)* written by Vitruvius circa 20 BC in Rome, and
Georgius Agricola's 1546 compilation of twelve volumes, *De re metallica*.
Vitruvius's work is an early instructional compendium of all technical
practices that were relevant at the time of its writing, from architectural
construction to hydraulic equipment like the pneumatically operated
musical organ to the construction of large-format chronometers and
assorted war machinery. Agricola—who, like Vitruvius, was equipped
with a universal polymathematical education—had aspirations no less
comprehensive. He was trained as a classical philologist, physician, and
pharmacist and was extremely knowledgeable in mineralogy and in mining techniques. On the basis of his knowledge of the latter, he developed
a demand—far ahead of his time, in this case—for a technical practice that could in the future be integrated with the humanist sciences.
"Moreover, the miner must be well versed in many arts and sciences:
philosophy first of all, in order that he recognize the origin, the cause
and qualities of subterranean things. For then he will have an easier and
more convenient way with quarrying and make better use of the ores he
mines."[6] Philosophy as a discipline that can be advantageous in the production and development of things external to human beings—the very
same assumption lies at the foundation of Kapp's frame of thought 331
years later.

Exactly 100 years before Kapp published his outline for a philosophy of technology, Johann Beckmann's *Anleitung zur Technologie, oder zur Kenntnis der Handwerke, Fabriken und Manufakturen* (Guide to technology, or the knowledge of handicrafts, factories and manufacturing [1777])[7] appeared in Vienna, which was then followed in 1780 by his *Entwurf der allgemeinen Technologie* (Design of general technology). In these two works, Beckmann introduces the concept of *Technologie* as a broad category encompassing the economy, social life, and the techniques of artifactual production. Earlier, Christian Wolff, in his 1710 book *Anfangs-Gründen Aller Mathematischen Wissenschaften* (Starting rationale for all mathematical sciences), had characterized *Technologie* as a special term to be distinguished from the more concrete *Technik*, which, in Kapp's sense of the term, signified the world of artifacts, of that which is artificially produced, of inventions objectified in materials.

In 1815, at the Berlin Book Printing Works, an extravagant pamphlet was produced that presented a unique variant on one such invention— namely, *Die eiserne Hand des tapferen deutschen Ritters Götz von Berlichingen* (The iron hand of the valiant German knight Götz von Berlichingen), which was prepared by Christian von Mechel on the occasion of the International Peace Congress in Vienna and was dedicated to the "crowned Emancipator of Europe."[8] Not only did Kapp later draw on this pamphlet conceptually, but he directly profited from it. In the chapter that deals with the hand ("Apparatuses and Instruments"), on page 102 of the 1877 German edition of his *Philosophie der Technik*, he point-blank plagiarizes the engraving of the German knight's iron prosthetic from Mechel's original, along with the detail of the mechanical finger's kinetics on page 103 (Figures 46 and 47). The illustrations command a veritably iconic status for his book.[9] Yet every element of symbolic and linguistic explanation in the extremely precise functional analysis of the artifact that Mechel undertook has vanished from the version of the engraving that Kapp pirates. That is why, in Mechel's description, all philosophical thought is lacking.

August Koelle's likewise vigorously anthropologically argued study *System der Technik* had already appeared by 1822, more than half a century ahead of Kapp's book. But Koelle's pioneering book was far less detailed and brilliantly written than Kapp's, and it lacked any provocative epistemological hypothesis that could compare with Kapp's organ projection.

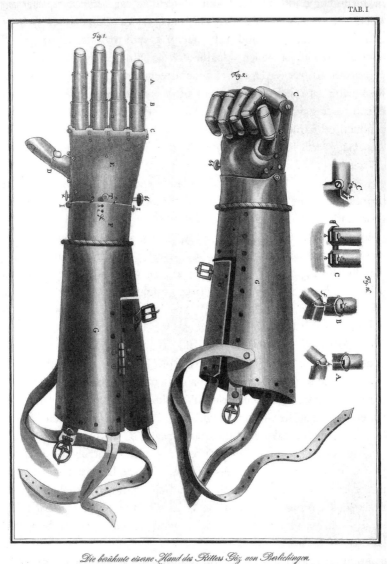

TAB. I

Die berühmte eiserne Hand des Ritters Göz von Berlichingen.
nach dem bey seiner Familie in Franchen aufbewahrten Original abgebildet von Chs. von Mechel.

Figures 46 and 47. The iron hand of the Knight of Berlichingen, taken from Christian von Mechels, *Die eiserne Hand des tapferen deutschen Ritters Götz von Berlichingen* (Berlin: Georg Decker, 1815).

TAB. II

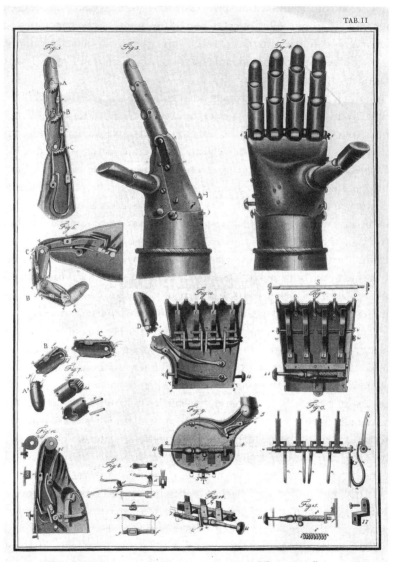

Götz von Berlichingens eiserne Hand, nach ihrem inneren Kunstreichen Mechanismus.
und allen zu demselben gehörenden einzelen Theilen richtig abgezeichnet

For this reason, it was not nearly as influential. Obviously, the same cannot be said of the foundational works of Karl Marx, in which he deduces
technology from the historical-materialist grounds of the conditions of
labor and production, in the context of his critique of political economy.
It is assumed that Kapp, a liberal and socially conscious intellectual, was
familiar with Marx's work.

Contemporary reality strongly reflected what Marx formulated theoretically and philosophically—even the everyday life that Kapp was living
of necessity in Texas, where he had emigrated in 1849, after the publication of his polemic on *Der constituierte Despotismus und die constitutionelle
Freiheit* (Constituted despotism and constitutional freedom), and where
he operated, among other things, a cotton farm. By the mid-nineteenth
century at the latest, more or less popular depictions of various equipment, instruments, and artifacts had begun to appear, representations the-
matizing the labor of the proletariat in factories or that of slaves in the
fields. Technology as the extension of the biological body even became
a theme in scientific magazines like the *Scientific American*. Hosford and
Avery's "Cotton Picker" (1858) can be read as the civil counterpart to the
militarily coded iron hand of the knight of Berlichingen (Figure 48).

The Philosophy of the Unconscious, Eduard von Hartmann's book that
Canguilhem refers to, first appeared in 1869, and several editions followed in rapid succession in Germany. The text ranks among the outstanding works that, long before Sigmund Freud, had begun to establish
psychology as a new academic and scientific discipline. The unconscious
as the ideational driver for the artificial construction of a world external
to the human psyche—which we have learned to describe as technology—
becomes an essential argumentative figure in Kapp's epistemological
construct of technology as organ projection. The figure can become particularly relevant for a perspective that thinks *techné* in its original unity
with art, the motivational power of which is not exhausted either in the
analogy to the organic nature of human beings or in the scale.

The nineteenth century is recognized as the age in which technology
was formulated as an essential field of operation for human beings and
their future, and it was during the nineteenth century that technology
evolved argumentatively in the most disparate disciplines. The imagination of human beings as interdependent functional correlations of biology, mechanics, hydraulics, and energy technology began to consolidate

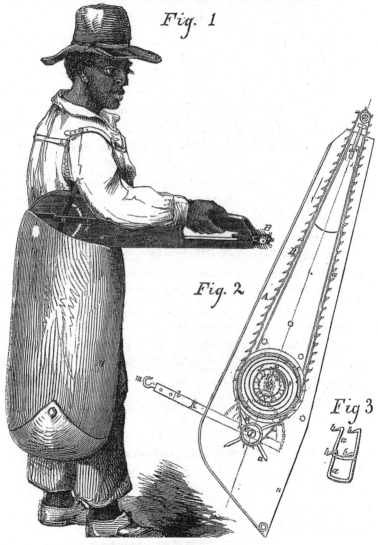

Figure 48. "Cotton Picker," taken from *Scientific American*, 1857.

and was increasingly purposefully integrated into the manufacture of apparatuses, equipment, and instruments of production. This process unfolded in parallel with industrialization and the capitalization of all conditions—social, political, and otherwise. It included the technical reifications that we would later categorize under the universalizing term *media*. Images, texts, and sounds became widely reproducible as technical artifacts and distributable around the globe as commodities. Photography, the zoopraxiscope and the bioscope, the panorama and diorama, the gramophone, telegraphy, film, telephony—Kapp was writing his book as the founding era of new media was unfolding around him, with new heights being achieved one after another in rapid succession. At massive World's Fairs, exhibitors and visitors celebrated the triumph of techno-rational thought over the ponderousness of the biological.[10]

To think technology as a *third element* that not only facilitates our reflections on the world surrounding us, aiding us in recognizing and mastering it, but also enables our self-recognition—this was the positive and progressive expression of the new Promethean conditions. Its inverse or negative expression—the production, via media of projection and reflection, of misrecognitions by an overwrought ego—would require decades yet before it was supposed. The figure of the toppling of technical reason over into its possible opposite would be philosophically thematized at length only after the experience of the Holocaust. Max Horkheimer described this qualitative reversal in his 1947 *Eclipse of Reason*. This foundational work of critical theory, which was likewise foundational for media studies as media critique in Europe, was originally written in English, and the German version did not appear until twenty years later.

SOME CONSEQUENCES

In the final decades of the nineteenth century, ideas and hypotheses like Kapp's had considerable repercussions. The establishment of technical colleges was of enormous significance for the advance of capitalism and the progress of various scientific-technical revolutions. These technical colleges came to replace polytechnical schools and the even older building and mining academies where students were taught technology, machine construction, engineering, and associated subjects. As Agricola had presaged, it was becoming clearer that, in order to evolve into an effective and hegemonic discourse that was paradigmatic for modernity, technology would require philosophy and psychology both, as well as law and economy, a poetic sensibility, and even profound philological thought. At my own alma mater, the Technical University of Berlin, but also for instance in Göttingen, students were still working even into the twentieth century in departments in which texts originally written in ancient Greek, Latin, Arabic, Persian, even Indian languages were systematically translated and annotated. It was not unusual to find textbooks containing in bilingual format—Arabic and German—Heron of Alexandria's writings on mechanics and robots, dating as far back as the first century. The aspiration to become a university was bound up in the European cultural context with an obligation toward the universalization of the idea of education itself. Philosophy was integrated into elite technical colleges first by means of lectureships and seminars, later through the establishment of faculties and institutes.

The fact that by 1913 direct monographic adaptations of Kapp's work began to appear—such as *Philosophie und Technik* by Eberhard Zschimmer, an engineer who held teaching posts in Jena, Karlsruhe, and elsewhere—indicates how significant Kapp's work must have been for this process. Zschimmer's critique of Kapp's theory of organ projection—focused above all on the metaphysical substance of organ projection, as a phenomenon steered by the unconscious—lags far behind the object of his polemic.[11] Although he shares with Zschimmer his criticism of Kapp's basic metaphysical assumption, Ernst Cassirer finds Kapp's project substantially more constructive for the prospect of a philosophy of technology from new perspectives, above all in his essay "Form and Technology": "This criticism does no harm to the basic idea and the fundamental insight that Kapp expresses when he writes that the technical operation, when outwardly directed, constitutes both man's self-revelation and, in him, a *medium* of his own self-knowledge" (emphasis added). Cassirer elegantly summarizes Kapp's basic thesis. In his *Philosophie der Technik*, Kapp "sought to think through the idea that the human being is granted knowledge of his organs only by detour through organ projection. By organ projection he understands the fact that the individual limbs of the human body do not simply operate externally, but they create an external existence, a quasi *image* of themselves. Each primitive tool is just such an image of the body; it is a counterplay and a reflection of the form and proportions of the body in a determined material construct of the outside world" (emphasis added).[12] The idea of organ projection as an act of production of an alternative reality external to that of the projecting body could hardly be more precisely summarized. Much later, Vilém Flusser, whose cultural anthropology was powerfully shaped by Cassirer's thought, would speak of the *new technical imagination* that is realized in acts of projection.[13]

In the 1920s, the Russian-Ukrainian avant-garde artist Solomon Nikritin (1898–1956) founded a group called "The Projectionists" based on a "theory of projectionism." The aim was to conceive a new system for a newly developing revolutionary social identity in the Soviet Union, which was to reach its apex in the creation of a human creative network. Nikritin composed and sketched fantastical analytical cartograms that could be classified as speculative-projective diagrammatic representations. In the place where Kapp's unconscious had been active, a new *consciousness* emerged that, "with the aid of the 'sciences and philosophies of art' and of the organization"[14] of the social, would be in a position to generate new projects of human creativity.

In the interwar period, which Sigfried Giedion termed the "era of total mechanization" in his 1948 *Mechanization Takes Command*, the relationships that Kapp had outlined had begun to reverse themselves. The by now well-elaborated system of mechanics served as a template for the description of the living, down to the finest biological detail. The arts, too, rapturously celebrated the "consummation of the human-machine alliance."[15] Mechanical as well as social and biological engineers were fascinated by the idea that the living body and life as a whole functioned effectively and could be manipulated and repaired just like an artifact or a technical system.

With direct recourse to Kapp, Fritz Kahn popularized this discourse in his five-volume compendium *Das Leben des Menschen* (The life of man), published in 1929, in which he fabricates a complete analogy between artificial and biological systems. He harks back to, among other things, the comparison between nerve strands and cable strands that originated with the nineteenth-century physician Rudolf Virchow and is now better known from Kapp's text, though it also makes an appearance in the writings of McLuhan. Still, what Kapp had already vehemently criticized a half century before Kahn now appears at the core of a projection thesis that completely inverts Kapp's epistemology. The activity of the heart, once seen as the machinal work of a system composed of chambers, conduits, and pumps, was modernized to become an "electro-technical system" operated by the "drive-apparatus of the heart," while the neuronal structure of the nervous system became a complex of conduits and switch relays.[16] The organic is no longer projected into the world external to the human being, but the reverse: the technical-mechanical serves as a template for describing and explaining the living. Between the two epistemological figures of excorporation and incorporation, thinking concerning the interface of the human being and the machine, of the media-human and the media-machine in particular, evolved over the course of decades.

KAPP: A MEDIA THEORIST?

Within the humanities, but also in mathematical and technical disciplines, *media theory* signifies a historically still rather young field of discourse. Texts attempting explicitly and systematically to provide theoretical frameworks or philosophical bases for consideration of medial phenomena only began to appear once diverse and particular technologies of *broadcasting* and even more so of *telematics* had begun to assert themselves as systems of mass communication. Philosophical dissertations like Max Bense's *Quantenmechanik und Daseinsrelativität* (Quantum mechanics and Dasein relativity [1938]) were no more seen as potential media theory than were, for instance, Walter Benjamin's artwork essay, Lacan's psychoanalytic concept of the mirror stage as ego formation (which he also developed in the 1930s), Ortega y Gasset's "meditations on technique,"[17] or Alan Turing's early essays on machines that proved themselves intelligent by virtue of the mathematician having assigned them the ability to learn.

Harold D. Lasswell—who, after publishing his dissertation *Propaganda Technique in the World War* in 1927, established himself as a leading figure in the study of modern mass democracies arising out of the spirit of technical and economic organizability—developed his renowned formula for describing any technically based communication with the five W's *(who says what through which channel to whom with what effect)* immediately after World War II, in the same year (1948) that George Orwell was formulating his bleak prognosis of a future televisually organized surveillance state in a novel that would become paradigmatic for the twentieth century. Lasswell's five-point formula later provided the theoretical structure for Lyotard's 1984–85 curated exhibition *Les Immatériaux*. Nearly parallel to Lasswell, Claude E. Shannon invented the information technology variant of a formula for technically based communication in his legendary contribution to the *Bell System Technical Journal* in July 1948, "A Mathematical Theory of Communication." Marshall McLuhan's 1951 *Mechanical Bride*—the title was a nod to Duchamp's *The Bride Stripped Bare by Her Bachelors, Even*—marks the beginning of a series of popular monographs in which the Canadian literary and communications scholar formulates his ethnological and anthropological observations on modern technocracies via the lens of mass media. Not only the idea that all technology is an extension of the human body—an idea whose spectacular formulation with Quentin Fiore catapulted McLuhan to stardom in the field of media theory—but also other of his core ideas like that of the global village today prove not original but derived. But McLuhan was able to formulate them admirably and thereby provide them a medial presence.

"If it is true that technology is our fate, then we must also consider it indispensable that technicians, in the interests of technical education, advance toward a philosophy of technology." Thus Johannes Erich Heyde (1892–1979), founder of a humanistic course of studies at the Technical University of Berlin after World War II, formulates the need for a humanities faculty with a strong philosophical emphasis. Heyde's successor, philosopher of science Kurt Hübner (1921–2013), goes a key step further, a step that contains the germ cell for the development of an explicit media studies curriculum. Hübner makes it unmistakably clear that the relationship between the humanist sciences and the natural sciences, like technology, must be a reciprocal one. Technical colleges and universities will "do better . . . at explaining the nexus of the humanities on the one hand and the natural sciences and technology on the other . . . than the philosophical faculties at universities of the purely classical type, who lack a living relationship with a reality that is by and large determined by technology. Consequently, philosophical faculties at technical colleges have the chance to create something new and to cope with today's urgent need."[18]

In fact, the Technical University of Berlin evolved along such lines to become a stunning generator of the new. During Hübner's term as chair of the philosophical faculty there (1960–71), a highly experimental Teaching and Research Institute for Language in the Age of Technology was set up under the leadership of Walter Höllerer. The institute was established in 1961, and by the end of the 1960s the first ever Department of

Media Studies in a German university had evolved under the leadership of Friedrich Knilli, an Austrian mechanical engineer, psychologist, and radio theorist.

A single institution that was capable of combining philosophy and technology, linguistics and poetics with electro-acoustics and electronic music, that could study particular media and their distinct materialities, and that could learn to research and organize both semiotics and aesthetics was unique in the postwar German situation and truly a sensation. It was taken for granted that here one would encounter authors and ideas like Kapp's *Philosophie der Technik* in the course of studying film and literature, radio or theater, and this encounter took place in close proximity with the philosophers of science and technology, the Latinists and the communications engineers that were on the faculty. Readings from the works of Kapp, Dessauer, Zschimmer, Ortega y Gasset, or Gehlen were just as much a part of everyday academic life as experiments with automatic translations, work with Karlheinz Stockhausen's advanced compositions, journals dealing with the reciprocity of the humanities and the natural sciences, and research conducted specifically through medial apparatuses like the tape recorder, the radio, or the video recorder. We, as young media researchers, were able to use all of these in order to develop our own ideas and beliefs regarding technology's penetration of cultural processes.[19]

Media theory, as a discipline, is historically bound to this short period in the long history of our civilization, in the course of which *the media* have become systemic and evolved their own ultimately hegemonic field of discourse. This has been the case as of the late 1960s and early 1970s. Kapp's thought, however, is much too broad-minded and far-reaching to be co-opted by such a discourse machine. Inside the highly charged interdependent relation between technoid and humanoid phenomena, questions and hypotheses have unfolded over more than two and a half thousand years regarding a magical mediating third—long before the media became systemic[20]—and these questions and hypotheses will continue to play a role long after an autonomous media discursivity loses its efficacy. Kapp is a bracing thinker of highly charged relationships like these, above all because he does not conceive of them in an apocalyptically debilitating way but as operational and forward-facing. The most important message in the context of his philosophical concept of organ projection lies not in his placing the beginning of the projection and manufacturing process in the unconscious. It lies rather in his thinking of psyche and material, of biology and structure, of the social and technology, of the particular and the general as indissolubly interwoven.[21] This thought arrives unabashedly ahead of its time for the last third of the nineteenth century, and it is not too late to radically reactivate it.

Yet, his message concerning the unconscious as that which animates the projection and the artifact is still of utmost importance and contemporary relevance—especially with respect to *techné* as a comprehensive experimental praxis that includes artistic production as well. Kapp's concept acquires the power of innovation via the gesture inherent to it of a propulsive dreaming, via the *Art of the Possible*[22] that Kapp so fascinatingly knows how to describe. It is for this quality that he has become over the past 140 years an important reference point in both the history and the philosophy of technology, which are not only deeply interwoven with one another but also with variants of a materiologically engaged media studies. I am convinced that, with this newly available translation, Kapp's ideas and concepts—like organ projection or the state as disciplinary machine comprised of parts functioning in circular full-closure—will enter and fortify the international field of media studies as well as, and more so, the more comprehensive field concerned with thinking the relationship of technology and civilization.

Berlin, June 2017

Notes

INTRODUCTION

1. Chris Olah et al., "The Building Blocks of Interpretability," 2018, Distill, https://distill.pub/2018/building-blocks/.

2. "Google Researchers Are Learning How Machines Learn," *New York Times*, March 6, 2018, https://www.nytimes.com/2018/03/06/technology/google-arti ficial-intelligence.html.

3. A major intervention on this topic was made by Harun Maye and Leander Scholz in their "Einleitung" to Ernst Kapp, *Grundlinien einer Philosophie der Technik: Zur Entstehungsgeschichte der Kultur aus neuen Gesichtspunkten*, eds. Harun Maye and Leander Scholz (Hamburg: Meiner, 2015), vii-l.

4. Page 10 in this volume.

5. Page 14 in this volume.

6. The notion that the human is by vocation technological—that there is no unmediated essence underlying the tool- and medium-maker "man"—is today most associated with the work of Bernard Stiegler, whose *Technics and Time* trilogy calls on Kapp for precisely this point. Bernard Stiegler, *Technics and Time I: The Fault of Epimetheus*, trans. Richard Beardsworth and George Collins (Stanford: Stanford University Press, 1998), 2.

7. Bernhard Siegert, *Cultural Techniques: Grids, Filters, Doors, and Other Articulations of the Real*, trans. Geoffrey Winthrop-Young (New York: Fordham University Press, 2015), 9.

8. Mark B. N. Hansen, *Embodying Technesis: Technology beyond Writing* (Ann Arbor: University of Michigan Press, 2000), 60.

9. Friedrich Kittler, *Eine Kulturgeschichte der Kulturwissenschaft* (Munich: Wilhelm Fink, 2000).

10. N. Katherine Hayles, *How We Became Posthuman: Virtual Bodies in Cybernetics, Literature, and Informatics* (Chicago: University of Chicago Press, 1999), 197.

11. Donna Haraway, "A Cyborg Manifesto: Science, Technology, and Socialist Feminism in the Late Twentieth Century," in *Simians, Cyborgs, and Women: The Reinvention of Nature* (New York: Routledge, 2010), 149–82.

12. Haraway, "A Cyborg Manifesto," 150.
13. Claude Lévi-Strauss, *The Raw and the Cooked*, vol. 1 of *Mythologiques*, trans. John and Doreen Weightman (Chicago: University of Chicago Press, 1983).
14. Bernhard Siegert, "Door Logic, or, the Materiality of the Symbolic: From Cultural Techniques to Cybernetic Machines," in Siegert, *Cultural Techniques*, 193.
15. Marshall McLuhan, "The Medium Is the Message," in *Understanding Media: The Extensions of Man* (Cambridge: MIT Press, 1994), 7. A number of scholars have connected Kapp to McLuhan's extension thesis, including Stefan Andriopoulos, *Ghostly Apparitions: German Idealism, the Gothic Novel, and Optical Media* (New York: Zone Books, 2013); and Jussi Parikka, *Insect Media: An Archaeology of Animals and Technology* (Minneapolis: University of Minnesota Press, 2010).
16. Friedrich Kittler addresses this legacy in *Optical Media: Berlin Lectures, 1999*, trans. Anthony Enns (Cambridge: Polity Press, 2010) and *Eine Kulturgeschichte der Kulturwissenschaft*.
17. Kittler, *Eine Kulturgeschichte der Kulturwissenschaft*, 204.
18. On this point we are grateful to Bryony Davies at the Freud Museum in London.
19. Kapp's programmatic manipulation of the flexible German word *Bild* proves difficult to maintain in English. The solutions we describe here preserve some but not all of the overtones of the original.
20. Page 53 in this volume.
21. Page 96 in this volume.
22. Karl Marx, *Capital: A Critique of Political Economy*, vol. 1, trans. Ben Fowkes (New York: Penguin, 1990), 286.
23. Hermann von Helmholtz, *Über die Erhaltung der Kraft: Eine physikalische Abhandlung* (Berlin: G. Reimer, 1847). On the general model that underlies these debates, see Anson Rabinbach, *The Human Motor: Energy, Fatigue, and the Origins of Modernity* (Berkeley: University of California Press, 1990).
24. Page 100 in this volume.
25. Pages 101–2 in this volume.
26. Page 102 in this volume.
27. Ludwig Feuerbach, *Das Wesen des Christentums* (Stuttgart: Philipp Reclam, 2008/1969), 31.
28. See Warren Breckman, *Marx, the Young Hegelians, and the Origins of Radical Social Theory* (Cambridge: Cambridge University Press, 1999), 90–131.
29. Breckman, *Marx*, 8.
30. Page 101 in this volume.
31. Günther Ropohl, "Karl Marx und die Technik," in *Die technikhistorische Forschung in Deutschland von 1800 bis zur Gegenwart*, ed. Wolfgang König and Helmuth Schneider (Kassel: Kassel University, 2007), 63–85; Guido Frison, "Technical and Technological Innovation in Marx," *History and Technology: An International Journal* 6, no. 4 (2008): 299–324.
32. Page 29 in this volume.
33. Page 29 in this volume.

34. Page 18 in this volume.

35. Herbert Spencer, *The Principles of Psychology* (London: Longman, Brown, Green, Longmans, 1855), 461.

36. For more about the reception of the extension thesis and the historicization of the senses with respect to instrumentation, see Christoph Hoffmann, *Unter Beobachtung: Naturforschung in der Zeit der Sinnesapparate* (Göttingen: Wallstein Verlag, 2006).

37. Page 106 in this volume.

38. Page 112 in this volume.

39. Page 111 in this volume.

40. Page 112 in this volume.

41. Page 112 in this volume.

42. G. W. F. Hegel, *Werke in 20 Bänden*, ed. Eva Moldenhauer and Karl Markus (Frankfurt a.M.: Suhrkamp, 1969–71), 7:24.

43. Kolakowski, in his magisterial work on Marxism, calls this "the Fichteanization of Hegel." See Leszek Kolakowski, *Main Currents in Marxism: The Founders, the Golden Age, the Breakdown*, trans. P. S. Falla (New York: Norton, 2005), 43–80.

44. His 1849 essay "Der constituierte Despotismus und die constitutionelle Freiheit" (Constituted despotism and constitutional freedom), on which the final chapter of *Elements* is in part based, brought Kapp the trouble with the authorities that led to his emigration. See Johannes Rohbeck, "Ernst Kapps Kulturetheorie der Technik," in *Materialismus und Spiritualismus: Philosophie und Wissenschaften nach 1848*, ed. Andreas Arndt and Walter Jaeschke (Hamburg: Meiner, 2000), 143–53, here 144.

45. See Leo Marx, *The Machine in the Garden: Technology and the Pastoral Ideal in America* (New York: Oxford, 2000).

46. Leander Scholz, "Der Weltgeist in Texas: Kultur und Technik bei Ernst Kapp," *Zeitschrift für Medien- und Kulturforschung* 1 (2013): 171–90, here 171.

47. Hegel, *Werke*, 10:303.

48. Hegel, *Werke*, 10:303.

49. Hegel, *Werke*, 10:303.

50. Page 23 in this volume.

51. See Wolf Lepenies, *Between Science and Literature: The Rise of Sociology* (Cambridge: Cambridge University Press, 1988).

52. Leo Marx, "'Technology': The Emergence of a Hazardous Concept," *Social Research* 64, no. 3 (1997): 561–77; Eric Schatzberg, "Technik Comes to America: Changing Meanings of Technology before 1930," *Technology and Culture* 47, no. 3 (2006), 486–512.

53. Hegel, *Werke*, 10:366.

54. Page 103 in this volume.

55. Page 110 in this volume. In his 1846 *Philosophical Geography*, which he refers to here, Kapp had developed a theory of "universal telegraphy" that is formulated

against the backdrop of Alexander von Humboldt's notion that the spread of science and experimentation would create "new organs" for the human. Kapp's book is a geography in the sense Johann Gottfried Herder had established: it tracks the influence of climate and environment on cultures, and then the advance of technological culture and its mutual influence on the environment. The final section, on "universal telegraphy," suggests that the early Kapp is following the movement that was called *Naturphilosophie* (of which his teacher Carl Ritter was a part). At the inception of that movement, Schelling had called for the "idealization of nature" and complementary "objectification of thought." Kapp seems to see their meeting point in telegraphy in 1846—and he is hardly the only Young Hegelian to draw on Schelling, who had posthumously become the nemesis of Hegelianism. But in *Philosophical Geography*, Kapp lacks the word *Technik* to describe the phenomenon he is pursuing. The interlude in the United States, combined with the vast extension of the telegraphic networks and the slow development of the word, mean that "universal telegraphy" can become subordinate to a full-scale theory of technology in 1877. But this also means that the philosophy of technology has sources in *Naturphilosophie* as well as in Hegel, something further confirmed by Kapp's constant recourse to the theory of the unconscious developed by Gustav Carus, whose twin works *Physis* and *Psyche* he draws on throughout *Elements*.

56. This syncs with Leif Weatherby's notion of "organology" in his *Transplanting the Metaphysical Organ: German Romanticism between Leibniz and Marx* (New York: Fordham University Press, 2016).

57. Friedrich Kittler, *Optical Media: Berlin Lectures, 1999*, trans. Anthony Enns (Cambridge: Polity Press, 2010), 29–30.

58. Bernhard Siegert, "Introduction: Cultural Techniques, or, the End of the Intellectual Postwar in German Media Theory," in Siegert, *Cultural Techniques*, 1.

59. Friedrich Kittler, "World-Breath: On Wagner's Media Technology," in *The Truth of the Technological World: Essays on the Genealogy of Presence*, trans. Erik Butler (Stanford: Stanford University Press, 2013), 122.

60. Kittler, "World-Breath," 2.

61. The claim "media determine our situation [*Medien bestimmen unsere Lage*]," which has been misunderstood almost as often as it has been cited, opens the preface to Friedrich Kittler's book *Gramophone, Film, Typewriter*, trans. Geoffrey Winthrop-Young and Michael Wutz (Stanford: Stanford University Press, 1999), xxxix.

62. Matthew Griffin, Susanne Herrmann, and Friedrich Kittler, "Technologies of Writing: Interview with Friedrich Kittler," *New Literary History* 27, no. 4 (Autumn 1996): 738.

63. Page 111 in this volume.

64. Hegel plays a prominent role in the second year of Lacan's seminar, which was also devoted to cybernetics, and on which Kittler draws heavily. Jacques Lacan, *The Seminar of Jacques Lacan, Book II: The Ego in Freud's Theory and*

in the Technique of Psychoanalysis, 1954–55, ed. Jacques-Alain Miller, trans. Sylvana Tomaselli (New York: Norton, 1991).

65. Ernst Cassirer, "Form und Technik," in *Symbol, Technik, Sprache: Aufsätze aus den Jahren 1927–1933,* ed. Ernst Wolfgang Orth and John Michael Krois (Hamburg: Meiner, 1985), 39–90, here 72–73.

66. Page 113 in this volume.

67. Pages 13 and 26; repeated on pages 53 and 161 in this volume.

68. Page 53 in this volume.

69. Page 53 in this volume.

70. Sigmund Freud, *Civilization and Its Discontents,* trans. James Strachey (New York: Norton, 1962), 38, 39.

71. Kittler, *Optical Media,* 30.

72. For more on the contest between experimental psychologists and philosophers in the German university system at the end of the nineteenth century see Mitchell G. Ash's *Gestalt Psychology in German Culture, 1890–1967: Holism and the Quest for Objectivity* (Cambridge: Cambridge University Press, 1998).

73. Page 10 in this volume.

74. Pages 10–11 in this volume.

75. Johann Friedrich Herbart, *Psychologie als Wissenshaft,* 2 vols. (Königsberg: Unzer, 1824, 1825).

76. Sigmund Freud, "Project for a Scientific Psychology" (1895), trans. James Strachey, *The Standard Edition of the Complete Works of Sigmund Freud,* vol. 1, *Pre-Psychoanalytic Publications and Unpublished Drafts, 1886–1899* (London: Vintage, 2001), 295–343.

77. Sigmund Freud, *The Interpretation of Dreams (Part I), The Standard Edition of the Complete Works of Sigmund Freud,* vol. 4, 1900, trans. James Strachey (London: Vintage, 2001) and *The Interpretation of Dreams (Part 2), The Standard Edition of the Complete Works of Sigmund Freud,* vol. 5, 1900–1901, trans. James Strachey (London: Vintage, 2001).

78. Kittler, *Eine Kulturgeschichte der Kulturwissenschaft,* 210.

79. Kittler, *Eine Kulturgeschichte der Kulturwissenschaft,* 210.

80. Sigmund Freud, *The Ego and the Id,* trans. James Strachey (New York: Norton, 1989), 20.

81. Page 63 in this volume.

82. Page 24 in this volume.

83. Page 24 in this volume.

84. Sándor Ferenczi, *Sex in Psycho-Analysis,* trans. Ernest Jones (Boston: Gorham Press, 1916). In general, the contested matter of projection in psychoanalysis dealt with schizophrenic or paranoiac breaches in ego boundaries that were repaired by "projecting" alien sensations outward in order to treat the source of the disruptive feelings as an external intruder. This was an attempt to shore up the ruptured border between the inside and outside. For more on this see

Jeffrey West Kirkwood, "The Cinema of Afflictions," *October* 159 (Winter 2017): 37–54. Opposite paranoiacs and schizophrenics who try to expel foreign sensations, the neurotic for Ferenczi "introjects" in an attempt to control as much of the outer world as possible. Although the descriptions of projection and introjection received the greatest attention in cases of pathology, Ferenczi notes in *Sex in Psycho-Analysis* that they were "merely extreme cases of psychical processes the primary forms of which are to be demonstrated in every normal being" (48).

85. Jacques Lacan, "The Mirror Stage as Formative of the *I* Function," in *Écrits*, trans. Bruce Fink (New York: Norton, 2006), 76, 77.

86. Siegert, *Cultural Techniques*, 11.

87. This preferentially organicist vocabulary in which certain terms or metaphors are irreducible to either side of a putative "mechanical/organic" divide resembles closely some aspects of the work of Georges Canguilhem, who cites Kapp prominently in his influential essay "Machine and Organism," in his *Knowledge of Life*, trans. Stefanos Geroulanos and Daniela Ginsburg, intro. Paolo Marrati and Todd Meyers (New York: Fordham University Press, 2008), 93–94.

88. Franz Reuleaux, *The Kinematics of Machinery: Outlines of a Theory of Machines*, trans. Alex B. W. Kennedy (London: Macmillan, 1876).

89. Page 122 in this volume.

90. Page 149 in this volume.

91. Page 149 in this volume.

92. Humberto R. Maturana and Francisco J. Varela, *Autopoiesis and Cognition: The Realization of the Living* (Boston: Reidel, 1980).

93. Page 163 in this volume.

94. Page 154 in this volume.

95. See Benjamin Peters, "Digital," in *Digital Keywords: A Vocabulary of Information Society and Culture*, ed. Peters (Princeton: Princeton University Press, 2016), 93–109, which construes the digital according to the differing functions of the hand.

96. Page 177 in this volume.

97. Pages 178–79 in this volume.

98. Page 179 in this volume.

99. Hegel, *Werke*, 8:196.

100. Hegel, *Werke*, 8:214–15.

101. Hegel, *Werke*, 8:235–36.

102. Gilbert Simondon, "On the Mode of Existence of Technical Objects," trans. Ninian Mellamphy, Dan Mellamphy, and Nandita Biswas Mellamphy, *Deleuze Studies* 5, no. 3 (2011): 407–24, here 411.

103. Pages 173–77 in this volume.

1. THE ANTHROPOLOGICAL SCALE

1. [The Grimm brothers' *Deutsches Wörterbuch* assumes the German *selb* to have derived from the compound form *si-liba*, "that which persists in itself,"

which Kapp extends into the expression *Leib und Leben*—body and life, or life and limb. It is possible that Kapp has borrowed the expression from Carus.—Trans.]

2. [The German *Person* may be defined philosophically as a being possessing a unifying form of consciousness by means of which it knows itself to be the same in spite of changing external circumstance.—Trans.]

3. Goethe, *Faust: A Tragedy,* trans. Walter Arndt (New York: Norton, 2001), lines 1335–36, 1338.

4. Meister Eckhart, from the tract "Sister Katrei," in *Ecstatic Confessions: The Heart of Mysticism,* ed. Martin Buber, trans. Esther Cameron (Syracuse, 1996), 153.

5. Adolf Lasson, *Meister Eckhart, der Mystiker: Zur Geschichte der religiösen Speculation in Deutschland* (Berlin, 1868), original modified.

6. Ludwig Feuerbach, *Das Wesen des Christenthums* (Leipzig, 1843), original modified.

[A general remark on Kapp's quotations and their translation: Kapp will occasionally alter his source materials when representing these for his own purposes. This may involve unremarked excisions and rearrangement of sentences, in part or whole. This passage from Feuerbach is a perfect case in point. Feuerbach's text reads as follows: "Als ein Spezimen dieser Philosophie nun, welche nicht die Substanz Spinozas, nicht das Ich Kants und Fichtes, nicht die absolute Identität Schellings, nicht den absoluten Geist Hegels, kurz, kein abstraktes, nur gedachtes oder eingebildetes, sondern ein wirkliches oder vielmehr das allerwirklichste Wesen, das wahre *Ens realissimum:* den Menschen, also das positivste Realprinzip zu ihrem Prinzip hat, welche den Gedanken aus seinem Gegenteil, aus dem Stoffe, dem Wesen, den Sinnen erzeugt . . ." Kapp's version: "Denn das Prinizip in [Feuerbachs] Philosophie, *'welche den Gedanken aus dem Stoffe, dem Wesen, den Sinnen, erzeugt, ist der Mensch, ein wirkliches oder vielmehr das allerwirklichste Wesen, das wahre* Ens realissimum,' im Sinne radikal materialistischer Schroffheit . . ." The present translation replicates quotations as they appear in Kapp's text. For this reason, most of the translations of quoted material are my own, even in cases where an English translation may already exist. For this reason, also, it has not always been possible to track down bibliographical information for passages where Kapp has not provided this himself.—Trans.]

7. Oscar Peschel, "Über den wissenschaftlichen Werth der Schädelmessungen," *Ausland* 45, no. 10 (1872), original modified; Edgar Quinet, *Die Schöpfung* (Leipzig, 1871), original modified.

8. Constantin Frantz, *Die Naturlehre des Staates als Grundlage aller Staatswissenschaft* (Leipzig and Heidelberg, 1870), original modified.

9. Karl Ernst von Baer, *Reden gehalten in wissenschaftlichen Versammlungen, und kleinere Aufsätze vermischten Inhalts* (St. Petersburg, 1876), original modified; Lazarus Geiger, *Zur Entwicklungsgeschichte der Menschheit* (Stuttgart, 1871), original modified; Heinrich Böhmer, *Geschichte der Entwicklung der naturwissenschaftlichen Weltanschauung in Deutschland* (Gotha: Besser, 1872), 210, original modified; Adolph Fick, *Die Welt als Vorstellung* (Würzburg, 1870), original modified.

10. Adolf Bastian, *Die Weltauffassung der Buddhisten* (Berlin, 1870), original modified.

11. Paul de Lagarde, *Über das Verhältnis des deutschen Staates zu Theologie, Kirche und Religion: Ein Versuch Nicht-Theologen zu orientieren* (Göttingen, 1873), original modified. [The phrase "e pur si muove" means "and yet it moves" and has been attributed to Galileo as his response to being suspected of heresy in 1633 by the Roman Inquisition for providing evidence of heliocentrism. The likely apocryphal quotation was first recorded in 1757 in Giuseppe Baretti's *The Italian Library.*— Eds.]

12. Carl von Rokitansky, in *Almanach der Wiener Akademie der Wissenschaften* (Wien, 1869).

13. Lazarus Geiger, *Ursprung und Entwicklung der menschlichen Sprache und Vernunft* (Stuttgart, 1868); Walter Bagehot, *Physics and Politics; or, Thoughts on the Applications of the Principles of "Natural Selection" and "Inheritance" to Political Society* (London, 1872); Carl du Prel, *Kampf ums Dasein am Himmel: Die Darwin'sche Formel nachgewiesen in der Mechanik der Sternenwelt* (Berlin, 1874).

14. Ernst Haeckel, *Anthropogenie: Keimes- und Stammes-geschichte des Menschen* (Leipzig, 1874), original modified.

15. Gustav Jäger, *Die Wunder der unsichtbaren Welt enthüllt durch das Mikroskop* (Berlin, 1868), original modified.

16. Pascal, *Pensées,* no. 347; Brunschwicg, original modified.

17. Ernst Kapp, *Vergleichende allgemeine Erdkunde in wissenschaftlicher Darstellung* (Braunschweig, 1869), original modified.

18. [The idea of a "system of needs" *(System der Bedürfnisse)* was first articulated by Hegel in his *Philosophy of Right.* Hegel's "system" fits together the individual's needs with the aggregate needs of a society in such manner that interdependence of need secures the satisfaction of need on every level. Kapp, so to speak, riffs on this theme, developing his own "system" to include such things as a "need to give form" *(Gestaltungsbedürfnis),* a "need for language" *(Sprachbedürfnis),* a "need of language for image" *(Bildbedürftigkeit der Sprache),* a "need of force" *(Kraftbedürfnis),* and so on. See chapter 6 for an elaborated definition of Kapp's "system of needs."—Trans.]

19. Julius Bergmann, *Philosophische Monatshefte* 5 (Berlin, 1870), original modified.

2. ORGAN PROJECTION

1. Transcript of the proceedings, *Der Gedanke: Philosophische Zeitschrift: Organ der Philosophischen Gesellschaft zu Berlin* (Berlin, 1861), original modified.

2. Carl von Rokitansky, *Der selbstständige Wert des Wissens* (Wien, 1869), original modified; Carl Gustav Carus, *Physis: Zur Geschichte des leiblichen Lebens* (Stuttgart, 1851), original modified; Karl Rosenkranz, *Hegel als deutscher Nationalphilosoph* (Leipzig, 1870).

3. Frederick Anton von Hartsen, *Grundzüge der Psychologie* (Berlin, 1874), original modified.

4. Friederich Ueberweg, "Zur Theorie der Richtung des Sehens," *Zeitschrift für rationelle Medicin* 5 (Leipzig and Heidelberg, 1859); Paul Kramer, "Anmerkungen zur Theorie der räumlichen Tiefenwahrnehmung" (1872); Carl Stumpf, *Über den psychologischen Ursprung der Raumvorstellungen* (Leipzig, 1873).

5. Carl Ludwig, *Lehrbuch der Physiologie des Menschen,* vol. 1 (Leipzig, 1858), original modified.

6. Johannes Müller, *Handbuch der Physiologie des Menschen für Vorlesungen,* vol. 2, 2nd ed. (Koblenz, 1840), original modified; Wilhelm Wundt, *Grundzüge der physiologischen Psychologie* (Leipzig, 1874), original modified; Adolf Horwicz, *Psychologische Analysen auf physiologischer Grundlage: Ein Versuch zur Neubegründung der Seelenlehre* (Halle, 1872), original modified.

7. [In 1868 the British biologist Thomas Henry Huxley believed he had discovered the *Urschleim,* the "primordial slime" thought to be the source of all organic life; he named it *Bathybius haeckelii,* in honor of the German scientist Ernst Haeckel—Trans.]

8. Karl Siegwart, *Das Alter des Menschengeschlechtes* (Berlin, 1874).

9. Otto Caspari, *Die Urgeschichte der Menschheit mit Rücksicht auf die natürliche Entwickelung des frühesten Geisteslebens,* vol. 1 (Leipzig, 1873), original modified.

3. THE FIRST TOOLS

1. Otto Caspari, *Die Urgeschichte der Menschheit mit Rücksicht auf die natürliche Entwickelung des frühesten Geisteslebens,* vol. 1 (Leipzig, 1873).

2. Lazarus Geiger, *Zur Entwicklungsgeschichte der Menscheit* (Stuttgart, 1871), original modified.

3. [*Werk-werkzeuge* in German. *Werkzeug*—"tool" in English—might be literally rendered as "work-thing" or "thing that works." Here the emphasis is on work in a material sense, i.e., these are tools that perform work. Later, Kapp will introduce tools of spirit, which do not "work" in quite the same way, but whose work is enabled by these earlier "working tools."—Trans.]

4. Geiger (1871).

5. Qtd. in Geiger (1871), original modified.

6. [Konewka was a nineteenth-century German illustrator who was known above all for his paper silhouettes.—Trans.]

7. Adolph Bastian, *Die Rechtsverhältnisse bei verschiedenen Völker der Erde: Ein Beitrag zur vergleichenden Ethnologie* (Berlin, 1872), original modified.

8. Georg Hermann von Meyer, *Die Statik und Mechanik des menschlichen Knochengerüstes* (Leipzig, 1873), original modified.

9. Ludimar Hermann, *Grundriss der Physiologie des Menschens* (Berlin, 1870).

10. Wilhelm Wundt, *Lehrbuch der Physiologie des Menschen* (Erlangen, 1865), original modified; Vierordt and Hermann qtd. in Wundt.

4. LIMBS AND MEASURE

1. Gustav Karsten, "Maß und Gewicht in alten und neuen Systemen," in *Sammlung gemeinverständlicher wissenschaftlicher Vorträge*, ed. Virchow and Holzendorff, vol. 6 (Berlin, 1871), original modified.

2. Conrad Hermann, "Gesetz der ästhetischen Harmonie und die Regel des Goldenen Schnittes," *Philosophische Monatshefte* 7 (Berlin, 1871/72), original modified.

3. Jules du Mesnil-Marigny, "The Economy of the Ancients," *Sonntagsblatt*, no. 11 (1873).

4. Wilhelm Förster, "Über Zeitmaße und ihre Verwaltung durch die Astronomie," in *Sammlung gemeinverständlicher wissenschaftlicher Vorträge*, ed. Virchow and Holzendorff, vol. 1 (Berlin, 1866), original modified.

5. APPARATUSES AND INSTRUMENTS

1. Friedrich von Hellwald, "Der vorgeschichtliche Mensch," in *Archiv der Anthropologie* 7 (Braunschweig, 1874), original modified.

2. Julius Zöllner, *Das Buch der Erfindungen*, vol. 2 (Leipzig, 1864), original modified.

3. Gustav Jäger, *Die Wunder der Unsichtbaren Welt enthüllt durch das Mikroskop* (Berlin, 1868), original modified.

4. Johannes Müller, *Grundriss der Physik* (1881); Ludimar Hermann, *Grundriss der Physiologie des Menschen*, 3rd ed. (Berlin, 1870); Carl Gustav Carus, *Physis: Zur Geschichte des leiblichen Lebens* (Stuttgart, 1851), original modified (here and below).

5. Hermann von Helmholtz, *Die Lehre von Tonempfindungen als physiologische Grundlage für die Theorie der Musik* (Braunschweig, 1863), original modified.

6. Johann Czermak, *Populäre physiologische Vorträge* (Wien, 1869), original modified.

7. Richard Hasenclever, *Die Grundzüge der esoterischen Harmonik des Altertums* (Köln, 1870), original modified.

8. F. J. Pisco, *Die Akustik der Neuzeit*, original modified.

9. William Thierry Preyer, *Die fünf Sinne des Menschen* (Leipzig, 1870), original modified.

10. L. Hermann (1870), original modified.

11. Zöllner (1864), original modified.

12. Pisco, original modified.

13. Czermak (1869), original modified.

14. Carus (1851), original modified.

15. Czermak (1869), original modified.

16. Czermak, *Das Herz und der Einfluß des Nervensystems auf dasselbe* (Leipzig, 1871), original modified.

17. *Das neue Blatt*, no. 35 (1873), original modified.

18. [Those terms suggestive of Kapp's influence on Heidegger's thinking of *Technik* are provided in square brackets in this text.—Trans.]

19. [In German: "das Werkzeug der Werkzeugung." Kapp is playing: *Zeugung* means to sire or to beget; *Werkzeugung* is his own invention.—Trans.]

20. [Kapp here refers to a political slogan, "Handicraft stands on gilded ground," which emerged in the context of contemporary political debates over industrialization.—Trans.]

21. Otto Caspari, *Die Urgeschichte der Menschheit mit Rücksicht auf die natürliche Entwickelung des frühesten Geisteslebens*, vol. 1 (Leipzig, 1873), original modified.

22. Alexander von Humboldt, *Kosmos: Entwurf einer physischen Weltbeschreibung*, vol. 2 (Stuttgart, 1847), original modified.

6. THE INNER ARCHITECTURE OF THE BONES

1. Karl Culmann, *Die graphische Statik* (Zurich, 1866).

2. Georg Hermann von Meyer, "Die Architektur der Spongiosa," in *Reichert und Du Bois-Reymonds Archiv* (1867).

3. Julius Wolff, "Über die innere Architektur der Knochen und ihre Bedeutung für die Frage vom Knochenwachstum," *Archiv für pathologische Anatomie und Physiologie* 50, no. 3 (1870), original modified.

4. Meyer, *Die Statik und Mechanik des menschlichen Knochengerüstes* (Leipzig, 1873).

5. *Der Naturforscher*, nos. 36/37 (1870).

6. [The prevailing understanding was that ossification of the bones took place differently in different locations, *in a way that is typical for the location*. Wolff is instead arguing that the difference is according to function rather than location.—Trans.]

7. Karl Rosenkranz, *Psychologie: Oder, die Wissenschaft vom subjektiven Geist* (Königsberg, 1843).

[This is an apparent affirmation of phrenology. For a discussion of the debate surrounding this "discipline" in the context of Hegel's *Phenomenology of Spirit*, see Lambros Kordelas, *Hegels kritische Analyse der Schädellehre Galls in der "Phänomenologie des Geistes"* (Königshausen & Neumann, 1998).—Trans.]

8. Goethe, "Typus": "Es ist nichts in der Haut, / Was nicht im Knochen ist."

9. Carl Gustav Carus, *Physis: Zur Geschichte des leiblichen Lebens* (Stuttgart, 1851), original modified.

10. Ludimar Hermann, *Grundriss der Physiologie des Menschen*, 3rd ed. (Berlin, 1870), original modified.

11. [See chapter 11, "The Fundamental Morphological Law," for an elaboration of *füg, Gefüge, Fügung,* and *Ineinsfügung* in relation to the coincidence of interiority and exteriority.—Trans.]

12. Max Heinze, "Die mechanistische und die teleologische Weltanschauung," *Grenzboten*, no. 42 (1874), original modified.

13. Paul Samt, *Die naturwissenschaftliche Methode in der Psychiatrie* (Berlin, 1874), original modified.

[The principle of *aktuelle Empirie* originates in the field of geology. The premise is that all forces that have acted on geological formations in the past must be considered valid and active in the present as well. The This also implies the same forces will be valid and active in the future, thereby enabling something like modeling or forecasting or, for Kapp specifically, the teleological view of development.—Trans.]

7. STEAM ENGINES AND RAIL LINES

1. [At no point does Kapp refer to a concept of "class"; rather, he speaks of *Stände*, or "estates." He elaborates on his concept of the estate in chapter 13.— Trans.]

2. Otto Liebmann, "Platonismus und Darwinismus," *Philosophische Monatshefte* 9 (1874), original modified.

3. Hermann von Helmholtz, "Über die Wechselwirkung der Naturkräfte und die darauf bezüglichen neuesten Ermittelungen der Physik," in *Populäre wissenschaftliche Vorträge*, vols. 1–3 (Braunschweig, 1876), original modified.

4. J. Robert Mayer, "Über die Ernährung," in *Naturwissenschaftliche Vorträge* (Stuttgart, 1871), original modified.

5. William Dwight Whitney, *Language and the Study of Language* (London, 1867).

6. Franz Perrot, *Vortrag über die notwendigen Schritte zur Hebung des deutschen Verkehrswesens.*

7. Max Perls, *Über die Bedeutung der pathologischen Anatomie und der pathologischen Institute* (Berlin, 1873).

8. Ludwig Feuerbach, *The Essence of Christianity*, trans. Marian Evans (London, 1854).

8. THE ELECTROMAGNETIC TELEGRAPH

1. Rudolf Virchow, *Über das Rückenmark* (Berlin, 1871), original modified.

2. Carl Gustav Carus, *Physis: Zur Geschichte des leiblichen Lebens* (Stuttgart, 1851), original modified.

3. Johann Czermak, *Das Herz und der Einfluß des Nervensystems auf dasselbe* (Leipzig, 1871), original modified.

4. Wilhelm Wundt, *Grundzüge der Physiologischen Psychologie* (Leipzig, 1874), original modified.

5. Phillip Spiller, *Gott im Lichte der Naturwissenschaften: Studien über Gott, Welt, Unsterblichkeit* (Berlin, 1873), and *Das Naturerkennen, nach seinen angeblichen und wirklichen Gränzen* (Berlin, 1873), originals modified.

6. Paul Samt, *Die naturwissenschaftliche Methode in der Psychiatrie* (Berlin, 1874).

7. Moritz Meyer, *Die Elektrizität in ihrer Anwendung auf praktische Medizin* (1868), original modified.

8. Emil du Bois-Reymond, *Untersuchungen über die tierische Elektrizität* (Berlin, 1848–60).

9. Alfred Dove, "Bekenntnis oder Bescheidung?" *Im neuen Reich* 2, no. 2 (1872), original modified.

[Fichte refers to the philosopher an "artist" who, when constructing a science of knowledge, is in effect inventing consciousness itself *after the fact."*—Trans.]

10. Hermann von Helmholtz, from the introduction to his German translation of John Tyndall's *Faraday as a Discoverer (Faraday und seine Entdeckungen)* (1872), original modified.

11. [The term *Gedankenwesen* (here rendered as "thought-entity") is Kant's designation for a product of pure thought, as opposed to actual objects that require our sensibility.—Trans.]

12. Eduard Reich, *Der Mensch und die Seele: Studien zur physiologischen und philosophischen Anthropologie und zur Physik des täglichen Lebens* (Berlin, 1872), original modified.

13. [This concept of a deficiency *(Mangel)*, the compensation for which is transformed to an advantage, is later taken up by Arnold Gehlen in his formulation of the human being as *Mängelswesen*. The advantage the human being has in his fundamental deficieny, according to Gehlen, is that he is then able to create a world of culture, which becomes his "second nature."—Trans.]

14. Ernst Kapp, "Universelle Telegraphik," in *Philosophische oder vergleichende allgemeine Erdkunde als wissenschaftliche Darstellung der Erdverhältnisse und des Menschenlebens nach ihren inneren Zusammenhang,* vol. 2 (Braunschweig, 1845).

9. THE UNCONSCIOUS

1. Eduard von Hartmann, *Philosophie des Unbewussten: Speculative Resultate nach inductiv-naturwissenschaftlicher Methode* (Berlin, 1868).

2. Carl Gustav Carus, *Psyche: Zur Entwicklungsgeschichte der Seele* (Pforzheim, 1846), original modified.

3. Karl Friedrich Zöllner, *Über die Natur der Cometen: Beiträge zur Geschichte und Theorie der Erkenntniss* (Leipzig, 1872).

4. Wilhelm Wundt, *Grundzüge der physiologischen Psychologie* (Leipzig, 1874), original modified.

5. Friedrich Zange, *Über das Fundament der Ethik: Eine kritische Untersuchung über Kants und Schopenhauers Moralprinzip* (Leipzig, 1872).

6. Rudolf Virchow, *Vier Reden über Leben und Kranksein* (Berlin, 1862), original modified.

10. MACHINE TECHNOLOGY

1. Franz Reuleaux, *Theoretische Kinematik: Grundzüge einer Theorie des Maschinenwesens* (Braunschweig, 1875).

[An English translation by Alex B. W. Kennedy was published in 1876 by Macmillan (London) as *The Kinematics of Machinery: Outlines of a Theory of Machines.* That translation has been consulted for terminology, but otherwise the translations here are my own.—Trans.]

2. [These "camps" that are opposed to one another here refer back to Greek antiquity; "aphoristic empiricism" is likely a reference to the style of medical

writings associated with the Asclepiads, an association of doctors that included Hippocrates.—Trans.]

3. [Kapp takes this statement from the opening remarks made by Virchow at a gathering of anthropologists, presumably in 1874 or 1875; see his review of Reuleaux's book in *Athenaeum* 1 (1875).—Trans.]

4. [The idea (still inchoate at this point) that force and motion in the machine is the counter-image of force and motion in human behavior will be more fully elaborated in the final chapter, "The State."—Trans.]

5. ["Common affection"—or *Gemeingefühl*—is noted in the Grimms' *Deutsches Wörterbuch* as "die sinnlichen Eindrücke oder Gefühle, die nicht unter einen der fünf Sinne zu bringen sind"; it is elsewhere referred to as a "mass of mixed sensibility" not included in the other classes of sensation.—Trans.]

6. [Kapp is again playing with *Werkzeug* and *Zeugung* here, *zeugen* being "to sire or beget offspring." See note 19 in chapter 5—Trans.]

7. [In the terminology of kinematics that Reuleaux in large part helped to establish, *Kraftmaschine* is often rendered in English as "prime mover," while *Arbeitsmaschine* is rendered as "direct actor." Here they are rendered more straightforwardly as "powering machine" and "working machine."—Trans.]

8. [The reference is to Robert Burns's 1795 poem "A Man's a Man for A' That." Written in support of the movement for Scottish independence, the poem sets the value of life as it is lived against the seeming worth of status conferred through mechanisms of power. Ferdinand Freiligrath translated Burns's poem into German in 1843 and published it in 1848 in Karl Marx's *Rheinische Zeitung*. Freiligrath's rendering of "for a' that, an a' that" as "trotz alledem und alledem" (which Kapp repeats here) appears again in the writings of Rosa Luxemburg and Karl Liebknecht.—Trans.]

9. [The German *Manufaktur*, like the English *manufactory*, predates industrial production and refers to handicraft and cottage industry, to manufacture by hand in general.—Trans.]

11. THE FUNDAMENTAL MORPHOLOGICAL LAW

1. [During a speech at the Berlin Academy of Sciences in 1880, Emil du Bois-Reymond introduces the phrase *"ignoramus et ignorabimus"* to distinguish between what science does not yet know and what science will never be able to know. The remark was met with accusations of "false modesty" by some.—Trans.]

2. Otto Caspari, *Grundprobleme der Erkenntnistätigkeit*, vol. 1 (Berlin, 1876).

3. Adolf Zeising, *Neue Lehre von den Proportionen des menschlichen Körpers, aus einem bisher unerkannt gebliebenen, die ganze Natur und Kunst durchdringenden morphologischen Grundgesetze* (Leipzig, 1854); *Das Normalverhältnis der chemischen und morphologischen Proportionen* (Leipzig, 1856).

4. Theodor Ludwig Wittstein, *Der goldene Schnitt und die Anwendung desselben in der Kunst* (Hannover, 1874).

5. Carl Schmitt, *Proportionsschlüssel: Neues System der Verhältnisse des menschlichen Körpers: Für bildende Künstler, Anatomen und Freunde der Naturwissenschaft* (Stuttgart, 1849); Carl Gustav Carus, *Die Proportionslehre der menschlichen Gestalt* (Leipzig, 1854).

6. Gustav Fechner, *Zur experimentalen Ästhetik* (Leipzig, 1871).

7. Conrad Hermann, "Das Gesetz der ästhetischen Harmonie und die Regel des goldenen Schnittes," *Philosophische Monatshefte* 7 (Berlin, 1871/72), original modified here and below.

8. Johannes Bochenek, *Die männliche und weibliche Normalgestalt nach einem neuen System* (Berlin, 1875), original modified below.

9. Wittstein (1874), original modified here and below. [See also note in chapter 8 concerning Fichte's definition of the philosopher as an "artist" who invents consciousness *after the fact.*—Trans.]

10. Robert Keil, *Corona Schröter: Eine Lebensskizze mit Beiträgen zur Geschichte der Genie-Periode* (Leipzig, 1875), original modified.

11. Wilhelm von Kügelgen, *Drei Vorlesungen über Kunst* (Bremen, 1842), original modified.

12. Qtd. in Emil Palleske, *Schillers Werk und Leben*, 4th ed. (Berlin, 1862), original modified.

13. David Friedrich Strauss, *Der alte und neue Glaube. Ein Bekenntnis* (Bonn, 1875), original modified.

14. Georg Hermann von Meyer, *Die Statik und Mechanink des menschlichen Knochengerüstes* (Leipzig, 1873), original modififed.

15. Zeising (1854), original modified here and below.

16. Bochenek (1875).

17. Zeising (1854).

18. Zeising (1854).

19. [Here Kapp is playing with a concept of measure, or *Maß*, that is literally at the root of proportion and symmetry, or *Gleichmaß* and *Ebenmaß*. The concept of *Maß* that progresses through *Gleichmaß* and *Ebenmaß* into harmony and from there into the quality of the beautiful in classical art is strictly distinguished, in the following paragraph, from a concept of *Maß* that is fixed into a quantitative rule or metric, as *Maßstab*.—Trans.]

20. Eduard von Hartmann, *Philosophie des Unbewussten: Speculative Resultate nach inductiv-naturwissenschaftlicher Methode* (Berlin, 1868), original modified.

21. Conrad Hermann, "Der wissenschaftliche Begriff der Aesthetik," *Philosophische Monatshefte* 5 (Berlin, 1870), original modified.

22. Zeising (1854).

23. [See chapter 4, "Limbs and Measure."—Trans.]

24. Zeising (1854).

25. Bochenek (1875).

26. Zeising (1854).

27. Qtd. in Zeising (1854), original modified.

28. [The Grimms' *Deutsches Wörterbuch* derives two particularly relevant meanings of "füge" (or "enge verbindung") from the following Latin phrases: *conjunctio vocum in cantando* (the conjunction of words in song) and *membrorum conjunctione* (the joining of limbs, or "gestalt"). Kapp's use of the terms *Gefüge, Fügung,* and *Ineinsfügung* should therefore be understood as strategically correlated with his elaboration of the Greek radical ἄρω.—Trans.]

29. Richard Hasenclever, *Die Grundzüge der esoterischen Harmonik des Altertums* (Köln, 1870), original modified here and below.

30. Albert von Thimus, *Die harmonikale Symbolik des Altertums* (Köln, 1868).

31. Hasenclever (1870).

32. [*Inzahl,* a term unique to Kapp, is rendered here as "innumeration," while *Anzahl,* rendered as "enumeration," is given a new and unique meaning in relation to the term *Inzahl.*—Trans.]

33. Julius Baumann, *Philosophie als Orientierung über die Welt* (Leipzig, 1872).

34. Lazarus Geiger, *Ursprung und Entwicklung der menschlichen Sprache und Vernunft* (Stuttgart, 1868), original modified.

35. [This progression of human capacities, from the coarse material to the intellectual, is based on a sequence of linguistically related terms: from *Naturgriff* to *Handgriffe, Kunstgriffe, Griff,* and finally *Begriff*; the condition of possibility for all of these capacities being the prehensile organ, or *Greiforgan.*—Trans.]

36. Zeising (1854).

37. Fechner (1871), original modified.

38. Gustav Fechner, *Vorschule der Ästhetik* (Leipzig, 1876).

39. Julius Zöllner, *Das Buch der Erfindungen,* vol. 2 (Leipzig, 1864), original modified.

40. Otto Caspari (1876), original modified.

41. Hermann (1871/72).

42. Otto Caspari, *Die Urgeschichte der Menschheit mit Rücksicht auf die natürliche Entwickelung des frühesten Geisteslebens,* vol. 1 (Leipzig, 1873).

43. Caspari (1873), original modified.

44. Moritz von Prittwitz, *Die Kunst reich zu werden, oder gemeinfaßliche Darstellung der Volkswirtschaft* (Mannheim, 1840).

45. Hermann Klencke, *Kosmetik, oder menschliche Verschönerungskunst auf Grundlage rationeller Gesundheitslehre* (Leipzig, 1869), original modified.

46. August von Eye, "Das Kunstgewerbe der Neuzeit," *Der Salon* 1 (1876), original modified.

47. Conrad Hermann (1871/72).

48. Friedrich Schiller, *Wallenstein* 2:6, original modified.

12. LANGUAGE

1. Ludwig Noiré, *Der monistische Gedanke: Eine Concordant der Philosophie Schopenhauers, Darwins, R. Mayers und L. Geigers* (1875).

2. Carl Gustav Carus, *Organon der Erkenntnis der Natur und des Geistes* (Leipzig, 1856).

3. Johann Czermak, *Populäre physiologische Vorträge* (Wien, 1869), original modified here and below.

4. W. D. Whitney, *Die Sprachwissenschaft*, ed. Julius Jolly (Munich: Ackermann, 1874); William Dwight Whitney, *Language and the Study of Language* (London, 1867).

5. Czermak (1869).

6. Max Müller, *Vorlesungen über die Wissenschaft der Sprache* (Leipzig, 1866), original modified.

7. Hermann von Helmholtz, *Die Lehre von Tonempfindungen als physiologische Grundlage für die Theorie der Musik* (Braunschweig, 1863), original modified.

8. [*Die Gartenlaube—Illustrirtes Familienblatt* was a weekly mass-circulation magazine, published in Leipzig by Ernst Keil from 1853 to 1878, when the publication changed hands. The editors of the 2015 Meiner edition of *Grundlinien* suggest that "Ott. B." may be Ottmar Beta.—Trans.]

9. Wilhelm Wundt, *Grundzüge der Physiologischen Psychologie* (Leipzig, 1874), original modified.

10. Heinrich Wuttke, *Geschichte der Schrift und des Schrifttums von den rohen Anfängen des Schreibens in der Tatuirung bis zur Legung elektromagnetischer Drähte* (Leipzig, 1872).

11. Lazarus Geiger, *Ursprung und Entwicklung der menschlichen Sprache und Vernunft* (Stuttgart, 1868), original modified here and below.

12. Conrad Kilian, *Im neuen Reich* 8 (1874), original modified.

13. Wilhelm Scherer, *Vorträge und Aufsätze zur Geschichte des geistigen Lebens in Deutschland und Österreich* (Berlin, 1874).

14. Friedrich Albert Lange, *Geschichte des Materialismus und Kritik seiner Bedeutung in der Gegenwart*, vol. 2 (1875), original modified.

15. *Scientific American* 31, no. 24 (1874).

16. Felix Dahn, "Wodan und Donar als Ausdruck des deutschen Volksgeistes," *Im neuen Reich* 8 (1872), original modified.

13. THE STATE

1. [*Gemeinwesen* has been rendered here as "commonwealth." The word *Wesen* is a motif that Kapp develops specifically in this final chapter. Because *Wesen* has no exact English correlate, I have not attempted to impose one on it, but have instead included the German term in brackets following my English rendering in the most relevant instances.—Trans.]

2. Adolph Bastian, *Die Rechtsverhältnisse bei verschiedenen Völkern der Erde: Ein Beitrag zur vergleichenden Ethnologie* (Berlin, 1872), original modified.

3. Rudolf Virchow, *Vier Reden über Leben und Kranksein* (Berlin, 1862), original modified.

4. Max Perls, *Über die Bedeutung der pathologischen Anatomie und der pathologischen Institute* (Berlin, 1873), original modified.

5. A. F. Grohmann, *Soziales Wissen* (Berlin, 1875), original modified.

6. Walther Flemming, *Über die heutige Aufgabe des Mikroskops* (Rostock, 1872).

7. Otto Caspari, *Die Urgeschichte der Menschheit mit Rücksicht auf die natürliche Entwickelung des frühesten Geisteslebens*, vol. 1 (Leipzig, 1873).

8. Gustav Jäger, *Die Wunder der unsichtbaren Welt enthüllt durch das Mikroskop* (Berlin, 1868).

9. Ernst Haeckel, *Natürliche Schöpfungsgeschichte: Gemeinverständliche wissenschaftliche Vorträge über die Entwickelungslehre im allgemeinen und diejenige von Darwin, Goethe und Lamarck im besonderen* (1868); *Anthropogenie: Keimes- und Stammes-geschichte des Menschen* (Leipzig, 1874).

10. Eduard von Hartmann, *Philosophie des Unbewussten: Speculative Resultate nach inductiv-naturwissenschaftlicher Methode* (Berlin, 1868).

11. Ernst Kapp, "Der constituirte Despotismus und die constitutionelle Freiheit" (1849).

12. Carl Gustav Carus, *Physis: Zur Geschichte des leiblichen Lebens* (Stuttgart, 1851), original modified.

13. Albert Schäffle, *Bau und Leben des sozialen Körpers*, vol. 1 (Tübingen, 1875).

14. Haeckel (1874), original modified here and below.

15. Caspari (1873), original modified. [Kapp refers here to three estates— *Wehrstand* (the military estate), *Lehrstand* (the intellectual estate), and *Nährstand* (the nutritional estate, which includes both agricultural and industrial production). This tripartite idea of the estates appears much earlier: "gott hat drei heilig stende auff erden, den lehr-, wehr- und nehrstand" (Johannes Mathesius) or "wehr-, lehr-, nähr-stand, ieder stand hat sein eignes her in sich" (Friedrich von Logau) *(Deutsches Wörterbuch)*.—Trans.]

16. [See endnote 13 in chapter 6 on *aktuelle Empirie*.—Trans.]

17. Goethe, *Goethe's Faust*, trans. Walter Kaufmann (New York: Random House 1963), 1:5, lines 1972–79: "Es erben sich Gesetz' und Rechte wie eine ew'ge Krankeit fort."

18. Ernst Kapp, *Vergleichende allgemeine Erdkunde in wissenschaftlicher Darstellung* (Braunschweig, 1869), original modified.

19. [Upon unification in 1871, disputes arose throughout Germany concerning how to nationalize the railway system, which was not accomplished before the twentieth century.—Trans.]

20. [See chapter 10, "Machine Technology."—Trans.]

21. Franz Reuleaux, *Theoretische Kinematik: Grundzüge einer Theorie des Maschinenwesens* (Braunschweig, 1875), original modified.

AFTERWORD

1. Grégoire Chamayou's French translation of Kapp's book appeared in 2007 as *Principes d'une philosophie de la technique* (Paris: Vrin).

2. Georges Canguilhem, "Machine and Organism"; English versions are available from Zone Books (*Zone 6: Incorporations*, 1992) and Fordham University Press (*Knowledge of Life*, 2008).

3. See above all the text by Friedrich Dessauer (1881–1963), "Technology in Its Proper Sphere," who published his own *Philosophie der Technik: Das Problem der Realisierung* (Bonn: Cohen, 1927); Dessauer simply took the title from Kapp fifty years down the line. Carl Mitcham and Robert Mackey's *Philosophy and Technology: Readings in the Philosophical Problems of Technology* first appeared in 1972 from Free Press; a number of later editions with thematic changes have come out since.

4. Chapter 3 in Grigenti's book is titled "Ernst Kapp—Organ Projection" (Dordrecht: Springer, 2016), 21–28.

5. Quoted in Johannes Rohbeck, "Ernst Kapps Kulturtheorie der Technik," in *Materialismus und Spiritualismus: Philosophie und Wissenschaften nach 1848*, ed. Andreas Arndt and Walter Jaeschke (Hamburg: Meiner, 2000), 143–44.

6. VDI (Verein Deutscher Ingenieure, Association of German Engineers) version of the works of Agricola, 1928, Book I, p. 1. Baseler version, p. 11: "Metallicus præterea sit oportet multarum artium & disciplinarum non ignarus: Primo Philosophiæ, ut subterraneorum ortus & causas, naturasque noscat: Nam ad fodiendas uenas faciliore & commodiore uia perueniet, & ex effosis uberiores capiet fructus." I owe an enormous debt to Hans Poser, who for many years occupied the chair of philosophy at the Technical University of Berlin; he generously permitted me access to his manuscript of the lecture "Philosophie an technischen Lehranstalten: Hintergründe am Beispiel Berlins," which he gave in November 2016 and from which this quotation is taken.

7. The complete title, with Beckmann's original orthography: *Anleitung zur Technologie, oder zur Kentniß der Handwerke, Fabriken und Manufacturen, vornehmlich derer, die mit der Landwirthschaft, Polizey und Cameralwissenschaft in nächster Verbindung stehn: Nebst Beyträgen zur Kunstgeschichte.*

8. *Die eiserne Hand des tapferen deutschen Ritters Götz von Berlichingen* (Berlin: Georg Decker, 1815), cited title signet.

9. The newly typeset reprinting of Kapp's book, with an introduction by the media scholar Stefan Rieger (Berlin: Christian A. Bachmann Verlag, 2015), even featured a fragment from Mechel's pamphlet on the cover. The first reprint of Kapp's book appeared in facsimile edition in 1978, with an introduction by Hans-Martin Sass. Why Felix Meiner published another reprint in 2015, even though Bachmann's reprint was only just recently published again, is beyond me.

10. In my book *Audiovisionen* (1989) I examined more closely the founding era of new media for "Rowohlt's Encyclopedia." The book was published in English by Amsterdam University Press in 1999, with the title *Audiovisions: Cinema and Television as Entr'actes in History*. In it, I make explicit mention for the first time of my own study of Kapp, also available in English (pp. 28 and 72).

11. Eberhard Zschimmer, *Philosophie der Technik: Vom Sinn der Technik und Kritik des Unsinns über die Technik* (Jena: Volksbuchhandlung, 1913).

12. Cassirer's "Form und Technik" appeared in 1930 and has been reprinted in various places, including Cassirer, *Gesammelte Werke*, Hamburg edition, ed. Birgit Recki, vol. 17, *Aufsätze und kleine Schriften (1927–1931)* (Hamburg: Meiner, 2004), 139–83. It is available in English in *Ernst Cassirer on Form and Technology*, ed. Aud Sissel and Ingvild Folkvord (Palgrave Macmillan, 2012), 15–53.

13. See *FLUSSERIANA—An Intellectual Toolbox* (Minneapolis: Univocal, 2015), a very compact edition of Flusser's world of ideas edited by Peter Weibel and myself, along with Daniel Irrgang.

14. See Lubov Pchelkina's entry on "Projectionism" in *AnArchive(s)*, based on an idea by Siegfried Zielinski, edited by Claudia Giannetti, compiled by Eckhard Fürlus (Oldenburg: Edith-Russ-Haus für Medienkunst, 2014), 136–37.

15. Lucie Schauer, "Vom Mythos zur Megamaschine—Zur Geschichte des künstlichen Menschen," in *Maschinen-Menschen,* exhibition catalog of the neue Gesellschaft für bildende Kunst (Berlin: nGbK, 1989), 7–8.

16. Fritz Kahn, *Das Leben des Menschen: Eine vollständige Anatomie, Biologie, Physiologie und Entwicklungsgeschichte des Menschen,* vol. 4 (Stuttgart: Kosmos, 1929), esp. 72–73 and 79. I have discussed Kahn and Kapp together in my short essay ". . . oder die Macht der Metapher," in *Lab—Jahrbuch 1996/7 für Künste und Apparate,* ed. Kunsthochschule für Medien Köln with the Verein der Freunde der KHM (Cologne: Walther König, 1997), 71–86.

17. Ortega y Gasset's *Meditación de la téchnica*, with a chapter on "Life as the Manufacture of Itself—Technique and Wishes," first appeared in Buenos Aires in 1939. It contains the sweeping remark, in the tradition of Kapp: "The original mission of technology is to give man the freedom to be himself." Quoted from the German edition, *Betrachtungen über die Technik: Der Intellektuelle und der Andere* (Stuttgart: DVA, 1949), 59.

18. Kurt Hübner, *Sinn und Aufgaben philosophischer Fakultäten an Technischen Hochschulen* (Berlin: Schriftenreihe Stifterverband für die Deutsche Wissenschaft, 1966), 48, qtd. in Poser 2016, 13–14.

19. I have begun to reconstruct this development in two different projects: in the monograph [. . . *After the Media]: News from the Slow-Fading Twentieth Century* (Berlin: Merve, 2011/English edition Minneapolis: Univocal, 2013) and in the project "Atlas zur Genealogie des MedienDenkens" (Atlas toward a genealogy of media-thinking), the first manifestations of which were an exhibition at HKW Berlin and the experimental *Atlas of Media-Thinking and Media-Acting in Berlin,* ed. Kristin Moellering and Leon Strauch in collaboration with Stefanie Rau (Berlin: transmediale, 2016).

20. In *Deep Time of the Media* (Rowohlt: Reinbek bei Hamburg, 2002/English edition Cambridge: MIT Press, 2006), I outlined the early beginnings of such systematization, with emphasis on the Greek pre-Socratics Empedocles and Democritus.

21. Henning Schmidgen puts ideas like these in a contemporary context alongside Deleuze / Guattari; see *Das Unbewusste der Maschinen—Konzeptionen des Psychischen bei Guattari, Deleuze und Lacan* (Munich: Fink, 1997), esp. 166–67.

22. Or, *Die Kunst des Möglichen*, as Christoph Hubig's three-volume work is titled, the first volume of which appeared in 2006, subtitled *Technikphilospohie als Reflexion der Medialität* (Bielefeld: Transcript).

Index

All page numbers in italics refer to illustrations.

observation of, 9; state of, 32; transfiguration of, 254

Naturphilosophie, 123, 270n55

nerves, 90, 104, 158, 187; auditory, 71; cross section of, *105;* electrical behavior of, 109; electrical wires and, 103; ganglion-free, 107; optical, 107; peripheral, 106

nervous system, xi, xxvi, 108, 112, 263; telegraphy and, xxvi, 103, 104, 106

New German Media Theory, xxix, xxx

New Theory of the Proportions of the Human Body (Zeising), 152

Newton's laws, xxxvii

Nikritin, Solomon, 262

Noiré, Ludwig, 203

noses, 91, *169*

"Notes toward a Theory of Spatial Depth Perception" (Kramer), 29

numbers, 56, 178, 179, 180, 196

numeration, 166, 167, 181

nutrition, 89, 97, 107, 160, 161, 208, 234, 284n15

nuts, 52; bolts and, 134; screws and, 133, 139

obedience, 241, 242, 243

object, 125; mechanical, 179; operation and, xi

"Odin and Thor as the Expression of the Spirit of the German People" (Dahn), 219

Oidtmann, Heinrich, 101

On a Method of Expressing by Signs the Action of Machinery (Babbage), 139

On Experimental Aesthetics (Fechner), 184

"On Nutrition" (Mayer), 97

"On the Architecture of Cancellous Bone" (Meyer), 82

"On the Inner Architecture of the Bones and Its Importance to Bone Growth" (Wolff), 82

On the Interaction of Natural Forces (Helmholtz), xix, 97

"On the Physiology of the Nerve Centers" (von Hartmann), 225

On the Psychological Origins of Representations of Space (Stumpf), 29

"On the Spinal Cord" (Virchow), 103-4

ophthalmoscope, 91, 110

Optical Media (Kittler), xxxi

organic, 52, 72, 103, 165, 228, 231; concept of, 98, 229; domain of, 109; mechanical and, 75; principles of, 181; sphere of, 246

organic activity, 50, 51, 72, 126, 146, 202

organic development, 17, 33, 89, 239; theory of, 32, 99, 118, 160

organic forms, 98, 102, 166, 227

organicism, xl, 230, 234

organic life, 17, 75, 81, 231, 245; collective, 19; knowledge of, 73; process of, 26

organic projection, x, 53, 81, 108, 232, 248; process of, 9, 101, 212; products of, 79

organic ratio, 164, 185; fundamental, 183, 187, 194; organ projection and, 189

organic structure, 73, 76, 91, 230, 244

organic vitality, 112, 228, 237, 246

organism, xix, 24, 49, 79, 90, 92, 99, 100, 102, 148, 186, 202, 209, 212, 230, 233, 234, 235, 236; antithesis of, 245; biological, 224; bodily, 7, 12, 23, 26, 41, 69, 96, 121, 187, 204, 223, 226, 228, 229, 248; collapse of, 240; concept of, 89, 112, 179; disclosure of, 111; function of, 247; growth of, 98, 222; healthy, 225; human, 219, 221, 226, 244; image of, 222, 226–27; individual, 19, 223, 233, 237, 245,

(continued from page ii)

ERNST KAPP (1808–1896) was a German philosopher of technology and a geographer. He was prosecuted for sedition in the late 1840s and subsequently immigrated to Texas, where he became a noted early freethinker and abolitionist.

JEFFREY WEST KIRKWOOD is assistant professor of art history at Binghamton University, State University of New York.

LEIF WEATHERBY is associate professor of German at New York University.

LAUREN K. WOLFE is a translator in the Department of Comparative Literature at New York University.

SIEGFRIED ZIELINSKI is Michel Foucault Chair at the European Graduate School. He is the author of *Deep Time of the Media: Toward an Archaeology of Hearing and Seeing by Technical Means* and *[. . . After the Media]: News from the Slow-Fading Twentieth Century* (Univocal, Minnesota, 2013).